The Ecstatic Experience

"Continuing the pioneering research of the late anthropologist Felicitas Goodman, her close friend and fellow researcher Belinda Gore presents twenty sacred postures that induce and facilitate the experience of religious ecstatic trance. Those who long for ecstasy have in this book a rich resource of simple strategies and proven ritual to achieve and satisfy that human craving."

JOHN J. PILCH, PH.D., PROFESSOR AT
GEORGETOWN UNIVERSITY AND
INTERNATIONAL SOCIETY FOR
SHAMANISTIC RESEARCH,
BUDAPEST, HUNGARY

"The work of Felicitas Goodman provides the best proof yet of the actual existence of an alternate reality and our hardwiring to connect with it. Belinda Gore continues Goodman's work, throwing open a door that's been sealed too long and showing the way back to this earliest of living traditions that was once common practice among ancient peoples."

LAURA LEE, HOST OF
THE LAURA LEE SHOW

"In 1977, the anthropologist Felicitas Goodman discovered that our bodies are hard-wired to access certain states of consciousness when we assume specific postures while being stimulated by rattling or drumming. She revived a ritual tradition that has been used for more than 35,000 years. *The Ecstatic Experience* is the first significant new research on sacred body postures since the death of Goodman in 2005. Belinda Gore has carried on the work of Dr. Goodman by doing continual research on new postures as they've been discovered, and now she offers 20 more ritual body postures that give us more ancient secrets that unlock the ancient wisdom hidden right in our own bodies. Long ago, these postures were a form of global communication that predates writing. As a library of humankind's emergence, they are a direct link to the archaic mind, and working with them could inspire us to create a healthy, imaginative world again. I recommend this book because by studying it and using the CD for rattling and drumming, you can discover the ancient codes of healing and transformation."

BARBARA HAND CLOW, AUTHOR OF
*THE MAYAN CODE: TIME ACCELERATION
AND AWAKENING THE WORLD MIND*

The Ecstatic Experience

Healing Postures for Spirit Journeys

Belinda Gore

Illustrations by Susan Josephson

Bear & Company
Rochester, Vermont

Bear & Company
One Park Street
Rochester, Vermont 05767
www.BearandCompanyBooks.com

Bear & Company is a division of Inner Traditions International

Library of Congress Cataloging-in-Publication Data
Gore, Belinda.
   The ecstatic experience : healing postures for spirit journeys / Belinda Gore ;
illustrations by Susan Josephson.
      p. cm.
   Includes bibliographical references.
   ISBN 978-1-59143-096-4 (pbk.)
   1. Trance. 2. Ecstasy. 3. Posture—Miscellanea. 4. Shamanism. I. Title.

   BF1045.A48G68 2009
   154.4—dc22

                                                                2008053687

Printed and bound in the United States by Lake Book Manufacturing

10 9 8 7 6 5 4 3 2 1

Text design and layout by Jon Desautels
This book was typeset in Garamond with Bergell used as the display typeface

To send correspondence to the author of this book, mail a first-class letter to the
author c/o Inner Traditions • Bear & Company, One Park Street, Rochester, VT
05767, and we will forward the communication.

To my mother, Sara,
who throughout her life was an
inspiring teacher

Contents

Acknowledgments

Without the wise guidance and encouragement of Barbara Hand Clow, this book would never have been born. In 1994 she took me under her wing and led me through the process of publishing *Ecstatic Body Postures: An Alternate Reality Workbook*. At the time Barbara and her husband, Gerry, owned Bear and Company Publishers and it was a great gift that they helped me birth my first book. Some years later, Barbara and Gerry sold Bear and Company to Inner Traditions, and thankfully Jon Graham was an enthusiastic editor who welcomed my new manuscript. I have had the pleasure of working with Laura Schlivek and the team at Inner Traditions • Bear & Company, and appreciate their support and professionalism that have made this book better than I could have imagined.

Throughout the book I have referred to individuals who came to monthly postures groups at my home and others who attended workshops both at The Cuyamungue Institute and around the United States. To protect their privacy, I have changed some names in the text, but I want to thank them all, especially Explorer Group members Merry, John, Pam, Jennifer, Sharon, another John, Michele, Bruce, Dominic, Jewelene, Geoff, Olga, and Ruth.

The Cuyamungue Institute continues to thrive thanks to a wonderful board of directors: Rae LeCompte, Jackie Haworth, Stephanie Stephens, John Pilch, and Nancy Sharp. We rely on our administrative director, Frances Wilson, to look after the Institute, handle workshop registrations, fill orders from the online store, welcome visitors, and thousands of other tasks small and large. It is a challenging role and thankfully over the years there have been outstanding people who have nurtured the land and the fledgling Institute. I can never express enough gratitude to Judy Morse, Linda Schroth, Carol Lang, and Joan Scott for their years of service.

Along the journey, I met the love of my life, who shares the ups and downs, and who knows to take me sailing when I have been too long at the computer. Thank you, John, for everything you are to me.

Introduction

This book has been gestating for over a decade. Weeks after the publication of *Ecstatic Body Postures: An Alternate Reality Workbook*[1] in 1995, I began gathering new postures to investigate during ecstatic trance workshops and ongoing groups. My intention was to organize them with the material that had been emerging to help us understand this wonderful and elusive quality we call ecstasy. Over the years, I would seem to be finishing the manuscript when a new insight would surface and I would begin again. When I think of the changes in my understanding, both of the impact of the use of sacred postures as well as the practices for sustaining expansion of consciousness, I realize how necessary this long gestation period has been.

Then on March 30, 2005, Felicitas Goodman, my friend and teacher, died. Now that she is no longer around for questions and conversation about our shared work, I appreciate even more her great wisdom and insight as a scholar, and I miss her. It is a tribute to her courage that throughout the ninety-one years of her life she was always willing to explore new territory, both as an anthropologist and as a woman, and she had the genius to recognize what she found. Her personal story about the discovery of sacred postures and her initial research about their use are told in her book *Where the Spirits Ride the Wind*.[2] Her study of the use of ritual postures in the spiritual practices of indigenous people around the world revived the practice of ecstatic trance at a time when the world is greatly in need of methods that can provide us with guidance and vitality. As she prepared to make her final journey to what she called the Alternate Reality, to live among her beloved spirit friends, she bequeathed the Cuyamungue Institute, which she founded as the center for education and research about the sacred postures, to those of us in the next generation. We are honored to carry on her legacy, and we are indebted to the hundreds of workshop participants with whom we have worked over the years.

Ecstatic trance has become a way of life for me, as natural now as breathing, but it took me a long time to recognize that it is a spiritual practice with many similarities to transcendent practices, both ecstatic and contemplative, throughout human history. I am not an anthropologist, like Felicitas, but rather a psychologist and a spiritual practitioner interested in the needs of those of us who are in the midst of global changes unlike anything humans have ever experienced. In my work with individuals, groups, and organizations, I have recognized that the key elements in ecstatic trance include precisely those tools and practices that can sustain us in these times of change. Medicine women and shamans around the world understood how to establish and maintain a relationship with the world of spirit. We can learn so much from them and, by combining their knowledge with our highly evolved left-brain thinking, can create a bridge to a new future for humanity.

For nearly three decades, Felicitas's research revived a tradition—at least thirty-six thousand years old—that uses the capacity of the human nervous system to alter its functioning very precisely in order to enter expanded or nonordinary states of consciousness. Using a collection of ritual body postures from the artwork of hunter-gatherer and horticultural people as a doorway to the world of spirit, this state known as ecstatic trance is achieved through a relatively simple, safe, and teachable method. It is evidence of the amazing durability of these sacred postures that groups of urbanized and technologically sophisticated women and men can assume the same body positions shown in the artwork of Paleolithic fishermen or Uzbekistani shamans and find themselves journeying into nonordinary states of consciousness, presumably in ways similar to these unknown ancestors.

Other anthropologists have recognized the potential in carvings and pottery figures as instructional pieces. In his last book, *The Active Side of Infinity,* Carlos Castaneda wrote, "some of the archeological pieces . . . had the capacity to produce . . . a shift of the assemblage point."[3] *The assemblage point* is his term for a Toltec concept that describes a perceptual organizing point, oddly enough in the upper back, that structures the way we receive input from the surrounding world and make sense of it. He was suggesting that some pieces of artwork found by archaeologists have such potent energy that they can induce a trance or nonordinary state of consciousness.

Another way to get our mental arms around this phenomenon of sacred postures is to consider them as codes. Rupert Sheldrake's theory of morphogenic fields suggests that there are invisible fields of energy that organize all systems in nature.[4] By enacting a simple ritual that includes assuming a body position shown in an archaeological statue or drawing, we can enter into morphogenic resonance with that artwork. Our bodies assume the same pose and we alert our nervous systems to attune to more subtle energy so that we can align with the same field as that represented by the statue. These morphogenic fields were fed by our ancestors' ritual activity that was designed to help them live in harmonious relationship with the spirit world. The sacred postures may be considered codes for finding similar ways to live in balance with the Alternate Reality.

During her last days, after several strokes had left Felicitas mostly unable to talk, I would read to her from her book *My Last Forty Days: A Visionary Journey Among the Pueblo Spirits.*[5] In the fall of 1995 she had told me that she was working on a new book but she could not tell me what it was about because it was a secret. Her eyes sparkled with a mischievous gleam as she told me that the book was a special gift given to her by the spirits. As it turned out, the story is a tale about her own death, a visionary story about the traditional forty days in Pueblo spiritual teachings between the time when someone dies and when she becomes a spirit. Written with her characteristic humor and wonderful storytelling style, Felicitas portrayed the process that she expected to undergo when the time came for her to die. After the strokes, I knew that it would soon be time for her to actually take that journey.

As I sat on her bed in the nursing home, reading to her, I would occasionally pause. "This is a good story, isn't it?" I would say. She would smile and nod. "Do you remember that *you* are the one who wrote this story?" I asked. No, she did not remember, but it pleased her that she had actually written this book that she was enjoying so much. Interspersed among the times spent reading were simple, straightforward conversations about what would happen when she died and would finally become a spirit herself. Since we knew how to make the journey to the Realm of the Dead using the sacred posture just for that purpose, I reminded her that I could visit her there and that one day I would join her when it was my time to die.

I thought maybe I was overdoing it with talking about dying but given the brain damage from the strokes I wanted to be sure she remembered how to make the journey to the Alternate Reality. Then one day, after some confused efforts to speak, she looked me straight in the eye and said quite clearly, "You know, Belinda, everything is the Alternate Reality." I smiled. She was quite clear about what was going on, and she had more wisdom about dying than I was giving her credit for. Yes, of course, she was right. *Everything* is spirit, and so there is no journey, really. We undertake spiritual practices to discover that reality is not fragmented or divided into different parts. It is a whole cloth, all around us. There are many paths, many practices for learning to live in the fullness of this reality all the time. Ecstasy describes the experience of knowing the world of spirit in the here and now, and living in it consciously.

I was with Felicitas about an hour before she died. She had been in a coma for some hours, and I wanted to say good-bye. As I sat with her it was her body that labored over the last breaths, but I had a vision of her dancing with her rattle in the New Mexican landscape that she loved. Months later I took her ashes to a special spot beyond our dance court at the institute and released them, in the place where, for her, the spirits ride the wind.

Ecstatic trance has been my primary spiritual practice for nearly twenty-five years, and I want to share with you the richness of this tradition as well as teach you the means to enter this expanded state of consciousness that Felicitas introduced to me. It is in honor of her wisdom, intellect, and great courage that I dedicate this book.

PART 1

Longing for Ecstasy

We have a collective longing for ecstasy, a hunger as fundamental and persistent as the need for food. How interesting that our bodies are designed—"hardwired"—for the experience of ecstasy and yet, for so many people in the contemporary world, the condition of ecstasy deprivation creates so much suffering. It was Felicitas's theory that ecstasy deprivation is the underlying cause of all addictions. As a psychologist who has treated alcoholism, eating disorders, and other addictive behaviors for many years, I wholeheartedly agree. Even though addictions are related to genetic predisposition and faulty neurology, the basic biology that produces the physical experience of ecstasy has gone haywire in a culture that does not teach us how to achieve it naturally.

Ecstasy is essentially a spiritual experience. We are ecstatic when our conscious awareness transcends the ego but at the same time aligns with the body, allowing us to be fully aware physically but without the inner dialogue of the mind. That is why sex is the form of ecstasy that many people can recognize. The physical experience of sexual pleasure overcomes the mind's incessant thinking and we are relieved, for the moment, of our brooding about the past and anticipation of the future. Among the ancient Egyptians there were specific rituals for using the ecstatic states awakened through sex to nourish the energy of the subtle body, known to the Egyptians as the *ka*. The hunger for ecstasy was acknowledged as a real hunger because ecstasy is food for the ka body, giving it vitality and potency. The ka not only sustains the physical cells and tissue, but also provides for the capacity to experience and express the conceptual reality, the Logos, that enlivens the physical tissue. Curiously, it is taught that shame is poison to the ka and that ecstasy is needed to detoxify the bodies from the negative effects of shame.

Other options for ecstasy are, of course, available. Drugs, alcohol, chocolate, and adrenaline rushes—from fear or extreme sports—can all provide the same initial experience but without long-lasting effect. Alternatively, native people around the world used to have a complex system of ritual body positions that make it easy to have an ecstatic experience. The use of a specific sacred pose accompanied by drumming or rattling can engage the body's natural ability to heighten brain activity and activate a state of consciousness that lies dormant during ordinary daily life. To experience that state is to experience ecstasy.

1

Living Ecstasy

The intensity of our longing for ecstasy has driven Western culture to be characterized by our addictions—turning to cigarettes, alcohol, food, work, material wealth, shopping, sex, and adrenaline highs—in search of a reliable source for ecstasy. Everything promises satisfaction, and advertising geniuses have learned to hook our deepest hunger for this satisfaction, but with solutions that never last very long. Brian Swimme, author of *The Universe Is a Green Dragon,* offers a challenge that "we need to confront the power of the advertised to promulgate a worldview, a mini-cosmology based on dissatisfaction and craving. Combining brain power and social status with sophisticated electronic graphics and penetrating psychological techniques . . . teams of highly trained adults descend upon children not yet in school with the simple goal to create in them—and in the rest of us—a dissatisfaction with their lives and a craving for yet another consumer product."[1] I might say instead that the dissatisfaction already exists. We remember a certain free abandon that we experienced as young children, a deep connection with our bodies and with subtle worlds. Advertising suggests that satisfaction is available again, through the consumption of yet another product. Even sex, that usually reliable source of ecstasy, has been commandeered to sell the products in glamorous ways, so that having an ordinary body and a simple loving relationship seems, well, ordinary.

We in the United States are technologically sophisticated enough to be able to reduce the working week by nearly 50 percent and still produce the goods and services that we need for a comfortable lifestyle. The problem is that we would have to make do without a lot of unnecessary stuff, and, more importantly, we would have to figure out what to do with all of our time. In one of the most materially and intellectually abundant cultures in the history of human existence, we are plagued with chronic and debilitating emptiness.

People living in less privileged circumstances have a hard time believing this could be true. When I was visiting the home of Russian tour guides in Moscow in 1991, my host caressed my Nikon camera and crooned, "You must be very happy people to have cameras such as these." I was embarrassed to see his raw desire and to realize how convinced he was that happiness lay in a finely engineered camera. But then I remembered that I also nurture illusions about potential bliss, just from other sources.

Our primary hunger is not for objects or the experiences they can provide, but for bliss and freedom. We are hungering for ecstasy and, ironically, ecstasy is an experience that requires emptiness. It calls for letting go rather than acquiring. When confronted with emptiness, we tend to indulge in our favorite compulsions, to fill up the emptiness with sensations and emotion, but we always have the option to explore the emptiness instead. Spiritual practices of every tradition, including those based on sacred postures, lead us into the emptiness. We must go through it in order to move beyond it. The longing for union and fullness draws us onto the path of return, giving us the motivation to face the challenge. In the practice of ecstatic trance we learn to invite the emptiness and then allow experiences beyond the familiar range of ego identity to open us up, to compel us to relinquish the imagined control of ordinary consciousness, and then to fill us with a sweetness and aliveness that exceeds whatever the material world alone can offer.

Leonard Shlain, in his remarkable book *The Alphabet Versus the Goddess,* suggests that the relatively recent introduction of reading and writing has markedly increased the development of the left hemisphere of the human brain.[2] The left brain is characterized by logical, rational, analytic thinking states that generate and rely on "hard facts." The right hemisphere tends to generate intuitive, innovative, emotional states that produce experiential learning, creativity, and interpersonal connection. The history of civilization has been the story of progressive rationality, precise measurement of time, future orientation, and written language, all specialized functions of the left side of our brains. To become so adept at these skills, on the whole we have sacrificed the talents of the right brain, which are intuition, emotion, creativity, appreciation of color and form, music, symbolism, and dream awareness. Although we are all capable of right brain activity, in general it is not highly valued in Western culture and economies. Consider how many young people have been advised to set aside a career in the arts for something more financially reliable, or how infrequently you hear about leaders in business and government consulting their intuition for guidance in decision making. Only recently have writers like Daniel Pink, in *The Whole New Mind,* and Daniel Goleman, in *Social Intelligence,* suggested that the scales are ready to tip in favor of the "softer" right brain qualities.[3]

Interestingly, to find enduring ecstasy we have to restore the power of the right brain without forfeiting the hard-won abilities of rationality. Right brain states are characterized by a sense of authenticity. "Once a person has experienced love or ecstasy, he or she *knows* it," writes Shlain. "The right hemisphere is also the portal leading to the world of the invisible. It is the realm of altered states of consciousness where faith and mystery rule over logic."[4] Another way to think about this is framed by Jungian analyst Marion Woodman. She writes, "Spirit without form is invisible; matter without spirit is dead. Matter and spirit love each other. They live through each other. . . . Their struggle to come together is the struggle toward consciousness."[5] In their struggle to come together, Woodman speaks of finding a hole in order to make it whole. The hole she speaks of is the dark abyss that inevitably shows up when we allow ourselves to give our full attention to the experience of emotion and body sensation in the present moment. Because of the need to encounter this hole, spiritual schools have traditionally required students to work with masters or teachers who can help them negotiate the dark passage they encounter. Our ancient ancestors, from whom we have inherited ritual body postures, were not so entrenched in the complexities of individualism and rationality. Nevertheless, even for them, the world of the spirits was sometimes considered dangerous. We now understand, thanks to the advantageous perspective of contemporary psychology, that the challenge is to maintain the balance of egoic identity while experiencing the world beyond ordinary consciousness, where we discover individuality to be an illusion. It can be both exhilarating and terrifying. However, once we have become accustomed to this expanded state of awareness, ecstasy is readily available.

Some years ago, I went to a writers' workshop, wanting help in getting this book started. The workshop facilitator for my group was a writer from New York City, a woman born in the Bronx who had cut her literary teeth writing political essays for *The Village Voice*. She was literally incapable of getting her mouth to form what was to her a ridiculous New Age–sounding phrase, "ecstatic trance." To her credit, she worked with me around creating a structure for good writing and only occasionally raised an eyebrow about the content of my story. The piece I had written was about a mystical experience in the ruins of the Maya city of Tikal. It was full of exotic imagery describing my encounter in trance with a Jaguar Priest. I was proud of it. My writing teacher had been to Tikal, she told us during the group session, and frankly found it disturbing. I dreaded our one-to-one session, knowing that she was not impressed by my ecstatic journeys. She sat down, looked at me, and asked, "What is this story about?" Intimidated by her full attention, I paused. She had read the story and I knew she was not asking me for a summary. What *was* the story about? Well, I said, it is essentially a story about loneliness.

I understood in that moment that my attraction to ecstatic trance was a longing for the deep sense of connection and belonging that I remembered from my early childhood. My parents and I lived in a little white saltbox house with red shutters on Mann Avenue in a small town in Ohio. Just down the street was Central School, where my mother was a seventh-grade social studies teacher. While my mother got ready for a day at school and my father was already at work as the manager of a local Sohio station, Hidgie from down the street would scoop me out of my crib and take me home with her for the day. Later, when I was a little older, Edith Wright from a few doors up the street would look after me until my mother picked me up on her walk home from work. The Sottoway boys lived next door and the Anderson twins and their brother Kent lived on the other side of us. And I could always dance and sing with my imaginary friend, Judy. I bristle when people say that an only child is usually spoiled, but the truth is that I never wanted for playmates or attention.

We moved away from Mann Avenue when I was four years old, and I lost my innocent sense of belonging in my world. For many years, I thought that moving had somehow wounded me emotionally because I could not account for the great sense of loss and longing that would at times well up in me. Much later, I learned that loss is an inevitable outcome of growing up and developing a personality, or the personal ego. Everyone experiences a sense of separation from the rich source of spiritual connection that we know as young children. Social conditioning gives us a sense of individual identity and teaches us how to function in the external world. In Western societies, we have lost the initiation traditions that helped our indigenous ancestors return to the mysterious world of spirit even during adulthood.

The story about Tikal exposed the deep loneliness that I saw in all of us on that trip. I was not the only one hungering for adventure and special experiences. We all wanted, in our own ways, to confirm a sense of connection with the spirits. The glamour of the Maya ruins, doing ritual trance at sunrise in the Grand Plaza, made it exciting. Ultimately, however, I wanted to know that the spirits I encountered during trances were real, that they were not figments of my imagination, and that they would continue to be real even in my own quiet home in Ohio. Even with all of its potential diversions, the world is a lonely place without the knowledge of how to expand beyond the limits of the material world.

I am indebted to my skeptical teacher because she challenged me to cope with the reality of my own loneliness. Otherwise my writing would have been a cotton-candy version of a onetime vision in the exotic setting of Tikal. By facing my loneliness I encountered the emptiness that sat in my belly, really engaged it, and felt the power of the Jaguar Priest's loneliness as well, as he faced the destruction of his culture. The ecstasy was not in tromping through the jungle in the hours before dawn to watch the sun come up over the ruins with the sound of howler monkeys in the background. All of that was exciting. But I felt ecstatic when I made a connection with the Jaguar Priest and felt his loneliness across centuries so that I could participate in the emotion with him. I had expanded beyond my own personality worries and discomfort to become part of something so much larger than myself while fully conscious and experiencing it in my body. That is what I mean by ecstasy.

Of course, I regularly get lost in the challenges of daily living, but through the practice of using sacred postures I know how to return to the state of ecstasy. This experience may seem different from meditative training that focuses on connecting with the Void, the great emptiness and nothingness out of which all form emerges. Ultimately they are the same, the form and the formless. To value one over the other is to be caught in duality. In my experience, it is helpful to practice both.

Often I wonder why the use of sacred postures has gone out of general practice, but perhaps it is only that the time has come for us to draw upon this ancient resource once again. As one Native American teacher recently remarked, we are entering an era when there will be more elders on the planet than in any other time in history. Some of us are finding ourselves stepping into roles of leadership during these changing and sometimes troubled times. Leadership in this world is taking on new meaning when we consider the predictions that the top ten jobs in 2010 did not exist in 2004, and that today's school-age learner will have ten to fourteen jobs by age thirty-eight; and when we consider that this world is so computer oriented that if MySpace were a country, it would be the eleventh largest in the world, between Japan and Mexico.[6] Our purpose in this work with ecstatic body postures is not only to have extraordinary visions, but also to use the practice to change ourselves so that we might create a better world in the midst of all this innovation. I welcome you to this journey of discovery, as together we find practical and powerful answers to satisfy our need for guidance and our longing for ecstasy.

2

Defining Ecstatic Trance

When Felicitas Goodman discovered the phenomenon of ritual body postures, she was looking for a way to teach her anthropology students how to speak in tongues. Actually, she had already shown them the preliminary steps for entering what she—and other anthropologists—called the religious altered state of consciousness. During her fieldwork in the Yucatán state in Mexico, she had observed ministers in small apostolic churches teaching their congregations how to speak in tongues, assuring that the Holy Spirit had entered them and that they were "saved" and would go to heaven when they died. Returning to her students at Denison University in Ohio, she instructed them in the same steps she had identified through careful observation of innumerable church services. But something was missing. Her students went into some kind of trance state but there was no speaking in tongues, no visions, no mystical states. Felicitas concluded that the religious altered state of consciousness required a belief system or dogma to shape the trance state. She was disappointed, but that was that.

Several years later, her attention was directed to an article in the R.M. Bucke Memorial Society's Newsletter Review in which Canadian psychologist V. F. Emerson reported that the belief systems of various meditative disciplines correlated with the different poses each system employed for meditative practice.[1] The subtle functioning of the body—heartbeat, breathing, even the motility of the intestines—was affected by the change in body posture. In other words, mudras, or postures, shaped meditative experiences because changes in the body were affected by each distinctive positioning of hands, arms, legs, and head, and those physical changes altered inner visionary experiences. Felicitas considered that perhaps body positions could shape trance states into religious altered states of consciousness. It was worth a try.

Scanning ethnographic journals and books, she searched for unique body positioning among the photographs of indigenous tribes engaged in religious rituals and of artwork that seemed to be linked to their ritual activity. Selecting six postures, she experimented with them with the help of individual students and colleagues who served as subjects. The results were astounding. She later wrote in *Where the Spirits Ride the Wind* that, "guided by hitherto unnoticed traditional body postures, these 'subjects' of a social-science experiment had taken the step from the physical change of the trance to the experience of ecstasy, they had passed from the secular to the sacred."[2]

Her research continues through the Cuyamungue Institute near Santa Fe, New Mexico. More than eighty ritual postures have been identified as doorways to physiologically precise altered states of consciousness in which ordinary people have ecstatic experiences: seeing visions, undertaking spirit journeys, and, yes, sometimes speaking in tongues.

Felicitas Goodman's rediscovery of ritual postures from indigenous tribal people offers a link between the worldview of hunter-gatherers and horticultural people and the lifestyle and spiritual challenges of city dwellers and anyone who has been influenced by the urban representation of life—which includes nearly everyone on the planet. The ritual postures were not transmitted in an oral tradition from teacher to student over generations. They were simply documented in cave drawings, totem poles, terracotta figures, and pottery designs. We do not know how the people of a particular tribe may have used them, so the method we teach is derived from observing ministers from agricultural villages instruct their congregations in how to achieve the religious altered state of consciousness, believing (and our experience of the past twenty-five years holds that it is true) that all ecstatic techniques have a fundamental similarity. Those original observations have been woven together with spiritual practices from many other traditions, most specifically from the Pueblo people of the Rio Grande Valley, our neighbors at the Cuyamungue Institute.

It has been both our strength and our challenge that we have had no one to teach us a particular method for using the postures. Because we have learned from our trance experiences instead of from teachers in the human world, the training has been long, sometimes confusing, and even misguided at times. However, this divergence from the traditional training of a shaman has allowed for a very important characteristic of our work: we are not tied to the cosmology, the traditions, the gods or spirits of any single tribe or society. This practice is truly a global teaching. We do not offer ourselves as teachers of the medicine wheel or the sweat lodge, for instance, or of specific shamanistic healing methods or rituals for the changing of the seasons or the cycles of the moon. Nevertheless, elements of many native traditions find their way into individual trances during workshops and personal sessions. Sometimes we recognize these teachings and rituals right away, and sometimes not.

As a gifted anthropologist, Felicitas could often identify in our experiences, shared around the circle after a session using a ritual posture, stories from vastly different cultural traditions. It is not necessary to know the stories for journeys into Alternate Reality to be meaningful. Nevertheless, it is wonderfully enriching to have within our group teachers who are knowledgeable about specific traditional teachings. One year the archetypal story that emerged through a Masked Trance Dance in Germany included the Duck spirit, not a usual participant in our rituals. I learned from Felicitas that there are old European legends about the Duck being the psychopomp, the one who leads those who have died to their next abode in the spirit world. Years later I discovered a reference to the Egyptian goddess Isis referring to herself as a duck: speaking to her brother/husband Osiris, Isis calls herself "thy duck, thy sister Isis." Someone who encounters the Duck spirit in her trance can explore her own associations with ducks, and then we can add the European legends and Egyptian references to extend the meaning that the Duck contributes to the experience. Ultimately, the meaning is a composite of personal and collective associations that continue to deepen the experience of the individual and the group.

The remarkable cross-cultural versatility of ecstatic body postures derives from a profound insight. Every ecstatic experience requires a ritual and with the sacred postures, the body itself provides the ritual when the body assumes the necessary pose. When the body holds the posture and the rhythmic sound of a drum or rattle gives the nervous system the required cues, the ritual is set in place. We enter the portal identified by the posture to journey into the Otherworld, or the Alternate Reality, to commune with spirit and to experience what exists beyond the world of consensual reality. It is that simple.

The body is the common denominator that joins all humans who have ever existed on the planet. Your skeleton and musculature, your hormones and arteries are pretty much the same as those of your ancestors fifty thousand years ago. Without access to their language or belief structure, you can experience the same capacity to interact with those mysterious multiple layers of reality documented by shamans in their rock art and be changed by waves of energy that realign the body for healing. These bodies of ours are genetically equipped to enter and leave contact with the spirit world on cue. Contrary to the messages conveyed by later hierarchies of priests and priestesses, no special status is required for these spirit journeys and they are accomplished with amazing ease: no years of preparation, no fasting or weeks of isolation required.

Why haven't we known about this? The postures themselves are no secret. They adorn museums around the world and stare back at us from cave drawings, totem poles, pottery, and ethnographers' photographs.

The issue is that no one recognized that nonliterate people might pass along the keys to their spiritual heritage in the most ancient form of "writing," which is images. Their artworks were their religious texts. They showed us in great detail how to assume the necessary poses that would lead us to the gods. But in the eyes of scholars, immersed in a culture of individualism, these were either random creative expressions of independent artists or were simple "primitive" magic and could not possibly be instruction or guidance for more sophisticated people such as ourselves. Consider the well-known cave drawing of the shaman on the wall of the Lascaux cave in France. He appears to be lying on his back with a giant aurochs, a Paleolithic bull, standing over him. There have been speculations that it is hunter magic and that the aurochs has killed the hunter, but his erect penis suggests otherwise. Some anthropologists have suggested that the bull was from a different drawing unrelated to the shaman. Since wall space in ritual caves was limited, drawings from different times are crowded together and can be discerned only by testing the pigments. Another possibility is that the shaman experienced the aurochs during his spirit journey. The bird head on the shaman and the bird on the top of his staff lying beside him give us the clue that this posture is used for making spirit journeys to the Sky World, where birds serve as messengers. A key instruction for this posture is to keep the elbow of the left arm rigid. To clarify this point, the elbow is actually circled. How more precise and specific could the artist be in giving nonverbal directions? That circle speaks to us across twelve thousand years.

Shamanism is a term used loosely these days to identify the spiritual practices of tribal people. In their article "Studies on Shamanism," anthropologists Kiikala and Hoppal acknowledge the booming interest in shamanism.[3] They attribute it to the tendency of researchers to label as shamanism everything that might involve healing or a spirit guide. They suggest that in a homogenized world of globalization, ethnicity has become highly attractive and the rituals of urban shamanism add an element of cultural richness in the world of mass culture. In their view, there are three benefits to urban shamanism that make it so appealing: the practices offer good relaxation in a stressful urban lifestyle; the techniques support discovering and healing psychic or psychological imbalances; and the experience of community engendered by these experiences is immensely beneficial to overly individualized modern city dwellers. The generation of relaxation, psychological healing, and the experience of community are also functions of yoga classes, group therapy, and spiritual workshops or retreats.

Compare the characteristics of an urban shaman with this description of the Maya shaman from Linda Schele and David Freidel's book, *Forest of Kings*. "The shamans were and are public explainers, repositories of the stories and morals of thousands of years of village experience. Their power is intimate and personal, and in the ecstasy of prayer their charisma is unquestionable. They are the keepers of a very complicated world view encoded in special poetic language. We call such knowledge oral history, but in fact it is much more than history. It is an ongoing interpretation of daily life."[4] In this context, the role of the traditional, pre-urban shaman is to keep his community connected with the true reality or the world of spirit that underlies the mundane embodied experience of daily life. The shaman knew the traditions of his people and could help them in applying the teachings from the spirits to the events of their daily lives, a role similar to that of a priest. The difference is that the shaman would interact regularly with the Otherworld through ecstatic experiences and would be able to share those experiences in meaningful ways with the community, and on some occasions assist them in entering that same world.

Flexibility of structure can allow a cultural role to adapt and evolve to meet the needs of living people. Shamans of old would pray to the spirits and ancestors that had been known to their people for generations. They were the keepers of the society's worldview, linking the living representatives of the tribe with the past, helping them to preserve their most cherished understandings of the world. Urban shamans as described by Kiikala and Hoppal are more similar to psychotherapists.

Psychotherapists can interpret the events and challenges of individual lives into language that supports the meaning of those experiences in the context of the cultural worldview. However, psychotherapy and most modern religious practices generally do not include ecstatic experience or rely on direct communication with the spirits as their primary source of guidance and wisdom. In fact, if contemporary priests, ministers, or psychotherapists suggest that they are in regular personal contact with God or disembodied spirits, they may face investigations for mental instability! Nevertheless, in the practice of ecstatic trance we teach people to do just that, to make ecstatic experience part of their regular spiritual practice and to rely on direct experience with the world of spirits as a source of guidance and healing.

The practice of ecstatic trance is not limited to the characteristics previously defined as urban shamanism. Ritual postures used in the context of a simple method described in detail in the next chapter allow the human body to shift its organs of perception to be able to experience and interact with multiple layers of reality. When the postures are from a particular culture or when a specific set of traditions and teachings focus the collective experience with a posture, then we tune ourselves to that layer in multidimensional reality. In addition, the belief system of the individual or their current psychological state shapes the experience to fit the needs of that person. We are accessing the wisdom traditions of many indigenous cultures and, at the same time, dealing with our subjective state at the time of the ecstatic trance experience as well as with our psychological makeup.

By teaching participants to access the Alternate Reality for themselves, we bypass the requirement that the shaman make the journey on behalf of the one in need of healing or guidance. However, the value of a well-trained and experienced facilitator cannot be underrated. I intentionally use the term *facilitator* rather than *shaman*, believing that we misrepresent ourselves if we who have been raised and schooled in Western ways take on the name of *shaman*. It is rightly used only by those who are of indigenous lineage or who have lived and been trained for many years among indigenous peoples. My own heritage is European but the spirits who most call to me are those from Central America. I will never be a shaman in the Maya or Olmec traditions, but the spirits whom I know through their artwork and ritual postures are my teachers and comrades.

One of Felicitas's most influential contributions to the understanding of religion was to reveal the assumptions about the world of spirit that are culture-bound. She wrote extensively about the impact of culture on perceptions of the Alternate Reality and the construction of a religious worldview. Religions that originated in agricultural societies envision a world of duality, in which there are light and dark forces working at odds with each other. In early religions, predating Christianity and the other spiritual systems of agricultural societies, the spirits who lived in the Other Reality were considered powerful and worthy of respect, but they were not divided into good and evil. The split into light and dark, good and evil, in both the ordinary reality and the Other Reality has had its consequences. If we look objectively at the world, we see that there is a natural wholeness of both darkness and the light of day, natural cycles of birth and death and decay. If we project the dark side onto the devil or the feared other, then we are condemned to act out destructively as we attempt to fight it. In our work with ecstatic body postures, we teach respect for power and powerful forces. Because a grizzly bear is powerful does not mean it is evil, just that it must be treated with respect. We are not trying to bend its power to our purpose; we are simply trying to coexist.

The definition of *shamanism* has also been distorted by the general lack of distinction between the traditional male path of shamanism and the female path. In her book *The Woman in the Shaman's Body,* anthropologist Barbara Tedlock differentiates the characteristics of shamanism as practiced by women, recognizing that what has generally been documented are the more aggressive male aspects.[5] For women shamans, the world of spirit, the Alternate Reality as Felicitas called it, is not dangerous. Our relationship with spirit requires respect and the quality of reciprocity, or the ability to both receive the gifts of the spirits as well as to offer gifts in return. The most prized gift to the spirits is open-heartedness and willingness to live in partnership with them. In a Tewa prayer of the Pueblo people it is said, "with tired backs we bring you the gifts that you love," and then the request is made, "weave for us now a garment of brightness."[6] Women shamans do not need to prove their courage. They prove their love and commitment to living with spirit by celebrating rituals, living in community, and practicing alignment with spirit with every breath. The body is not sacrificed, but honored as the means by which spirit enters the material world.

The practice of ecstatic trance with ritual body postures is a path of the feminine side of shamanism. It is body-centered, accessing the Alternate Reality through simple postures and an easily learned ritual that is used to attune to a natural capacity that is part of everyone's genetic makeup. Because access is based on abilities that are hardwired into everyone's bodies, this method is not exclusive but available to all. It is practiced in groups and in community, and its use fosters a sense of commonality rather than hierarchy.

Contact with the spirit beings who reside in multidimensional realities is based on establishing relationship, similar to developing friendships among people. Reciprocity, or the give-and-take aspect of friendship, is preferred over subservience or worship. Through the experience known as metamorphosis, we humans are able to shape-shift and soften the boundaries that separate us from other forms of life. We can become a wolf or hawk or butterfly in the other worlds, rather than being stuck in just one species. In ritual dances, workshops that we call Masked Trance Dances, we humans make masks and cos-tumes of the animal spirits and offer them a chance to materialize in this world through us in ritual, similar to the animal dances of native tribes throughout the world.

There is nothing wrong with male shamanism. It is just not the only option, and we need to recognize the variability within shamanis-tic practices, both the masculine forms and the feminine forms found in cultures around the world. Ecstatic trance is simply one path among many. Through the practice of using ritual postures, gifts from our collective ancestors, we have the opportunity to learn a simple method for having direct experience of the world of spirit. We all bring the basic equipment—the human body—and with the guidance of trained teachers, we can all learn a safe and readily accessible way to make spiritual ecstasy a natural part of our lives.

3

Why We Use Ecstatic Body Postures

In workshops around the world, people ask me, "Why should we do trance? What's the purpose?" The first reason to undertake ecstatic trance is that it feels good. We know from research that in the course of a fifteen-minute ecstatic trance session, there is an increase in the production of beta-endorphins in the brain, resulting in an increase of these endorphins in the blood and therefore throughout the body.[1] Endorphins are naturally produced opiates and provide for physiological changes in the body that made opium such a popular drug: relaxation of tight muscles, release from obsessive thinking, and a general sense of well-being. These are the same sensations that accompany a satisfying meal and good sex. Felicitas often said that, during her fieldwork in the Yucatán in the hot Mexican summers, she could be relieved of the lung-tightening heat and humidity when she went into the trance state, and all the problems of living as an outsider in an unfamiliar culture seemed to dissolve after an evening of religious ecstasy in the little apostolic church where she was investigating glossolalia. When workshop participants complain that they have had no visions during trance, I remind them that at least they have fulfilled a very fundamental need to feel content and at peace with the world. Not a bad outcome for only fifteen minutes of effort.

A second purpose for ecstatic trance is to experience that expansion of consciousness that ecstasy provides. We need regular opportunities to be reminded of the magnificent resources our brains and bodies can provide if we only call on them. I expand my consciousness any time I function "outside the box," because I am literally extending the boundaries of what I can include in the space-time configuration I call my "self." Once I have stretched the perimeter of "me," there is more room to explore my own possibilities. Exploring is fun and can sometimes be an end in itself.

A third reason for undertaking journeys into nonordinary reality is to ask for specific guidance. The experience of insight in trance comes from two sources. Because I have expanded my conscious awareness of the resources that already exist within myself, I am able to access what Jungian dream worker Jeremy Taylor calls "the deep unconscious that contains all that waking consciousness desires and longs for the most—the energies of love, creativity, and felt communion with the Divine."[2] Jung's concept of the shadow is fairly widely known as the negative aspects of self that cannot be embraced and therefore are pushed into the unconscious, to be recognized only when we project those traits onto others, and judge the other instead of ourselves. But Taylor and Jung suggested that there are also positive, wise, and noble aspects of ourselves that we cannot quite acknowledge, out of fear that we will lose our identities, the little notions of self that we cling to. When we open up in trance, we can make connections with this bright shadow, this God within, and find solutions to problems and insights into perplexing problems.

In addition to the wisdom that comes from within is the very real experience of making contact with nonphysical beings that populate the Alternate Reality—call them spirits, angels, or ancestors. Carlos Castaneda called them "inorganic beings of awareness."[3] The postures by their very physical definition open quite specific doorways into the Alternate Reality, so that we are not wandering around crying for help to any discarnate being that we might encounter. When we use the divining postures in particular, we are going to a specific location (if I may be excused for using spatial language to describe a nonspatial reality) to request the presence of the spirit shown in the carving or figurine. That is, we use the Egyptian Diviner Posture to ask for guidance from the being we know as the Egyptian Diviner. We seek help quite literally from the Man from Tala by using the postures of the same name, and we call upon the Maya Oracle by assuming her specific pose.

Throughout the years I worked with Felicitas she was emphatic in explaining that these experiences we were having during ecstatic trance were not symbolic but real encounters with real spirit beings. She was concerned that we would fall into the prevailing cultural worldview that, in this psychological age, tends to relegate every internal experience to an abstract reflection of our own selves. Sometimes this is true, but not always. A good parallel is our dream experience. Within dreams, as the nervous system clears itself after a day of stimulation, we may experience jumbled impressions that reflect nothing more than the debris of the daily flood of stimuli. We can also have real dreams that deal symbolically with events from our daily lives. Another category of dreams—sometimes called Big Dreams—reflects a deeper, even mystical reality. Dreams in this last category are often not symbolic at all but are encounters with a reality beyond daily life. Similarly, in ecstatic trance we may see colors or shapes, or think about something that concerns us, but other times—most of the time—we have clear encounters with something beyond ordinary experience. We may need to explore the meaning of these trance events but translating the meaning does not reduce the event to "nothing but" a symbol. It is real.

A fourth reason for engaging in ecstatic trance is to seek healing, which from the perspective of this worldview means finding balance. Healing can and does occur at many levels, often at the same time. We know from psychosomatic research and mind-body medicine that humans are complex creatures composed of continually interacting systems. We cannot effectively compartmentalize disease to one dimension only. If a person comes to me for psychotherapy with emotional symptoms of depression—feeling hopeless, lethargic, and generally bad about himself—we must treat more than the obvious symptoms because depression is a multilevel disorder. He may have a sluggish nervous system that needs a boost with medication, and at the same time, he may require behavioral tools to help deal with his difficulty with concentration that is jeopardizing his job. Additionally, he may need to strengthen his support system through group involvement or through developing his skills for reaching out to the people in his life. We will have to explore possible problems in relationships, disappointments in his life, underlying anger, and a lack of connection with his spirituality and sense of meaning and purpose. Once we open to the possibilities of who we are through transcendent experience, we realize there is even more to be aligned and balanced for optimal homeostasis.

Finally, a fifth reason for undertaking ecstatic trance is to be of service in restoring the cosmic patterns that have been disrupted by human activity. Felicitas shared with us the story of a hunter who disregarded a taboo by killing too many deer and was sent to the other reality to make amends for his transgression. One function of trance, she believed, is to participate in rearranging the order of the world to fit the continual evolution of the manifest reality. The health of the universe, according to the people who lived on the earth before we became enamored with the power to control nature, is maintained by living according to patterns (call them archetypes if you want) that keep the world in order. Since we have the freedom to decide to function differently—the problem of free will—we can screw things up pretty badly, and environmentalists are easily able to give us examples of how we are continually doing so. We can amend the problems, however, and participate in giving form to a new world by participating in ecstatic trance.

4

The Basic Method for Entering Ecstatic Trance

Because it was her early goal to develop a simple and teachable method for entering religious ecstasy, Felicitas Goodman focused on establishing a repeatable structure that could exist independently of organized religion. She first identified this method while doing fieldwork in the small apostolic church in the Yucatán state in Mexico. Because ministers in apostolic churches needed to teach their parishioners how to speak in tongues in order to demonstrate that they had been possessed by the Holy Spirit, these ministers were expert in guiding people into the religious altered state of consciousness. While their instruction was described within the context of Christian doctrine, Felicitas recognized several steps that were consistently utilized. These steps were not dependent on beliefs but rather created a template based more on physiology and basic psychology. As I began to teach workshops, I saw the structure fill out more and more, and I realized that the elements of ecstatic trance are similar to those used in contemplative practices.

There are five fundamental steps in the practice of ecstatic trance by ordinary individuals who are not trained as shamans, but who have the same inherent capacity to interact with the world of spirit:

1. Preparing oneself
2. Establishing sacred space
3. Quieting the mind
4. Stimulating the nervous system
5. Undertaking a ritual, in this case through the use of a ritual posture

To fully understand each element of the practice, we need to review each step in more detail and explore ways to enhance the experience at every level.

Preparing Oneself

Preparation is essential because we have the ability to interfere with the shift into nonordinary states of consciousness if we choose to do so. It is this capacity to exert personal will that acts as a safety valve and allows us to teach this method with confidence, knowing that the body and mind of the practitioner can and will function to ensure personal security. Workshop participants often ask if everyone can do trance, and yes, almost everyone can. The few exceptions include people whose nervous systems, once stimulated, cannot respond to the organizing principle offered by the ritual, as is the case with people with psychotic disorders. However, normal people can also resist trance when they have trained themselves not to respond to stimulation. A good example is practitioners of Zen meditation.

In her early workshops in Vienna, teaching at a Zen center, Felicitas was surprised when enthusiastic participants had difficulty entering the state of ecstatic trance. Eventually they worked out the cause of the problem. These meditators had practiced for years to learn to empty their minds of every thought that arose. Zen Buddhist meditation is the ultimate contemplative practice in which the goal is emptiness. Concomitant with the practice is a calming of the nervous system, heart rate, and brain wave patterns. Ecstatic trance is not contemplative in nature but is the result of stimulation of the nervous system, with an increased heart rate, but with slower brain wave patterns. The Zen practitioners were feeling the stimulation and calming themselves. They would begin to see inner visions and would empty their minds. The two practices worked against each other. With an understanding of the differences and recognizing what was required for ecstatic trance, these workshop participants were soon able to learn to shift into a different mode than their accustomed meditative state.

Psychological preparation is as simple as anticipating the experience of trance with positive expectation: "This is beneficial to me and I will come away from the experience feeling better about myself and about the world." It is a state of mind that facilitates the shift of consciousness that we identify as the trance state. As soon as new practitioners learn how to use ritual body postures and they have a few journeys under their belts, we also teach clarifying an intention for the trance so the ritual posture can be selected. The choice of posture sets the intention. It is not the focus of mind that leads one into the Alternate Reality, but rather the position of the body. Ultimately, individuals recognize that they receive what they need, regardless of the posture chosen—but specific postures do support specific types of experiences. The individual or group decides whether to approach the spirits for healing, for guidance, or for a metamorphosis. Group participants can make short statements about the life issues that are their current focus, or when an individual is undertaking the trance alone, a few moments of quiet time or journal writing will accomplish the same purpose.

Fasting is often a preliminary to journeying in the world of spirits. However, during a workshop, I do not encourage people to fast, only to avoid eating heavy foods on the days of the workshops. Because many people eat diets of processed food, short-term fasting often functions as a method for cleansing the body to rid it of toxic chemicals. If someone is in the midst of detoxifying, this physiological process can interfere with the changes the body and nervous system need to undergo to enter into and maintain the trance state. Supervised fasting in preparation for a workshop can be helpful if people are willing to undertake this cleansing on their own, but it is not required.

During the ritual in preparation for ecstatic trance, we use smudging as a way to clean out mental and emotional cobwebs and to become psychologically ready to shift into an expanded state of consciousness. The practice of smudging has been revived with the current interest in Native American spirituality. In churches and temples, its counterpart is the use of incense. We burn a fragrant herb or resin to create smoke that cleanses both the space and each individual's personal space or energy field. Placing a stone in a large flat shell or fireproof pottery, you can burn a piece of charcoal (there are small, self-lighting briquettes sold in religious stores or where you might buy incense) and put a little bit of sage or cedar, or a chunk of resin, like copal, onto it. Drawing the smoke over your head and body, you are clearing away energetic stuffiness. The traditions say that smoke lives in both worlds, in the visible physical world because we can see it, but also in the spirit world because it has no substance.

Establishing Sacred Space

Once you are prepared, the next step is to establish sacred space. Churches and temples are the spaces traditionally used for spiritual rituals, but we can create sacred space anywhere using a few simple steps. T. S. Eliot wrote, "Except for the point, the still point, there would be no dance, and there is only the dance."[1] To create a still point in a turning world is to create a safe container for the experience of transcendence, a container that has well-defined boundaries and the protection from intrusion and interruptions. Choose your space well. I have tried to lead groups in ecstatic trance in outdoor settings—under the giant ceiba tree in the ruins of Tikal, at the temple to Aphaia on the Greek island of Aegina, on a grassy spot in Machu Picchu—but I have always been uncomfortable in doing so because we could not escape the curious eyes of other visitors to these sites. Participants rarely complained and often had wonderful visionary experiences, but I was uneasy and more than once called upon the Jaguar Spirit to protect us. In trance, I could watch her snarling around the perimeter of the group to guard the boundaries. In other situations, we have comfortably held workshops outside but they all took place within a prescribed area, like the dance court at the Cuyamungue Institute or deep in the Bear Cave at Blacktail Ranch in Montana.

It is important to create a beautiful space as well, perhaps with candles and flowers, which are also components of many religious services. An altar establishes the center focus and is the place to put all of the ritual materials, like a rattle, a smudge pot, and candles, as well as any special sacred objects that are meaningful to you.

Along with preparing the physical space, we also sanctify it by making an invocation to God or the gods or the Holy Spirit or holy spirits. In our practice we invite the spirits to be with us by first awakening the spirit of the rattle or drum we are using. A simple method is to take a pinch of cornmeal—tobacco or sage will do as well—and to blow onto it. My breath identifies that I am the one sending this gift to the spirit of my rattle. Then I make a circle around the rattle, giving just a bit of the gift to the east, south, west, and north of my rattle, and above it and below it. Felicitas taught me to say something like this: "Wake up, Little Sister, sound good and true." Sometimes I add, "Sing sweetly to the spirits and call them for us." Turning to the east, I say, "Spirits and grandparents in the east, the place where the sun rises and all things begin, join us here in our ritual. Come be with us now," and I shake the rattle four times. Turning to the south, I say, "Spirits and guardians of the south, the place of fullness, of the growing season and of midday, please join us now. Come, come be with us," and again I shake the rattle four times. Facing the west, I say, "Spirits, grandmothers, keepers of the heart, those who live in the going-within place, come join us in this ceremony. Come be with us," and I shake the rattle four times. And facing the north, I say, "Ancestors, spirits, keepers of the wisdom, living in the place where the cold comes from and where the seed lies buried deep within the soil, resting, we call on you and ask you to join us now," and again I shake the rattle four times. Finally, reaching my arms above me with the rattle in my hand, I call out, "Spirits of the Sky World, sun, moon, wind, and Milky Way, we invite you to join us here. Come be with us now," and I shake the rattle four times, and then bending toward the earth, I say, "Rise up dear spirits of the Earth, rise up and join us here. Come be with us in our ritual." Again I shake the rattle four times. Of course, the directional associations change when I am teaching in the southern hemisphere, but the model can be easily adapted.

Following any invocation, it is appropriate to offer a gift. In the Methodist churches where I grew up, that gift was a hymn or a prayer of praise. In the space to which we have called the spirits, we blow on another pinch of cornmeal, offer it to the four directions and above and below, and say, "To all of you who have heard our call and have come, thank you for your presence here, and welcome." Then we toss the cornmeal toward the sky, returning the gift of the Corn Mothers who, according to Pueblo tradition, gave their bodies so the people, who were starving, could eat.

Quieting the Mind

Step three in the process of ecstatic trance is a practice to quiet the mind. The internal chatter that is ever present in people with developed left hemispheres—that is, almost all of us who are able to read this page—effectively interferes in the shift of consciousness required for ecstasy. People have used various hallucinogens to suppress this function but that adds additional risk, both physically and legally. For a practice outside the context of a ritual guided by a well-seasoned religious specialist, we do not recommend the use of drugs or consciousness-altering substances. A simple alternative is to learn the fundamental meditative practice of emptying the mind, usually with a breathing technique.

A simple method is to sit comfortably and count your breaths. On the inhalation, focus on a spot just below your navel and witness the rising of your belly as you draw in a full breath. Then allow that breath to naturally release, that is, exhale from your belly, and count that as one breath. Pause, and then take in a second breath, expand your belly slowly and release, and count "two." Inevitably the mind begins to wander: "Is this the right way? What do I look like? Remember what happened the last time I tried to meditate? What was it that I wanted to remember to do this afternoon?" and on and on. When that happens, just bring the focus of your attention back to the breath and the counting. Slowly, the chatter dies down and our sense of presence in the moment deepens. Try counting fifty breaths prior to using a sacred posture for trance: counting ten breaths, holding down one finger, and repeating that four more times. This is also an effective technique anytime you need to calm your mind. I recommend it to my psychotherapy clients who have a difficult time going to sleep or who develop anticipatory anxiety.

Sometimes in a group people prefer to just focus on the breath and only the leader does the counting. I recommend counting because it gives the mind something to do and helps keep it from wandering.

Stimulating the Nervous System

The three steps identified so far are useful in preparation for meditation and contemplative practices as well as for ecstasy. The next step differentiates the two paths. While in meditation we learn to calm both mind and body, in the practice of ecstatic trance we need to stimulate the nervous system. Our method is to use a drum or rattle to create a steady even rhythm of about two hundred beats per minute, a beat that is consistent with ceremonial rhythms at the corn dances and animal dances of the Pueblo Indians who are our neighbors at the Cuyamungue Institute.

To initiate the change in consciousness that we know as ecstasy, you have to *do* something to the body. Little children induce mild states of ecstasy when they delight in spinning, again and again, on the merry-go-round, or in twirling their bodies until they fall. Clapping, dancing, chanting, and singing all have a similar effect. Harsher and more demanding stimulants like flagellation, long periods of deprivation, and various kinds of naturally occurring drugs are also effective and all have been traditionally used by shamans and other religious specialists to induce ecstatic transcendence. The beauty of working with sacred postures is that they are so profound that simple drumming or rattling is sufficient and cannot be misused with dangerous consequences. The sound affects the brain so that rational processes are subdued and other functions are stimulated. The sound of the rattle must be evenly consistent throughout a fifteen-minute session to facilitate the shift in nervous system functioning. Learning to rattle for trance sessions is part of our training for practitioners and instructors at the Institute. Given the precision of this means of stimulating the brain and nervous system, it is possible to enter the trance state on cue—with the beginning of the sound of the rattle accompanied by a ritual posture—and even more importantly to return from the altered state on cue. This latter capacity distinguishes intentional journeys into Alternate Reality from the random wanderings of people with psychotic disorders, who cannot choose when the shift in consciousness occurs, nor where the shift will take them.

Felicitas and I made a tape many years ago that includes four sessions, each totaling fifteen minutes of sound: the first session records the sound of one rattle (Felicitas), the second includes two rattles (both of us), the third records only a drum (me), and the fourth includes both a rattle and a drum (both of us). I prefer the resonance of both the drum and the rattle when live sound is not possible. Included at the back of this book is a CD of these four sound sessions. Using the CD, it is possible to put on earphones and use the sacred postures to achieve ecstatic trance by yourself. Or using speakers, a new group can practice together until someone learns the art of rattling. Opposite the CD sleeve at the back of the book, you will find further instructions for using the CD along with the ritual body postures.

Using a Ritual Posture

Simply stimulating the body, and more specifically the nervous system, will not provide a religious experience. Felicitas learned in her early research that given all four of the previous conditions, her subjects would certainly experience shifts in consciousness, but none had what might be called a spiritual experience or an ecstatic trance. Her preliminary conclusion was that it was not possible to teach people to experience ecstasy without a shared religious dogma to provide context and meaning; that is, she believed people could not be possessed by the Holy Spirit, as in her fieldwork on glossolalia, or call upon any other spirits, without having a well-structured belief system in place. The body alone seemed insufficient to produce ecstasy. However, thanks to a serendipitous discovery of an article about the possible impact of using mudras, or postures, during meditation, she proved herself wrong. The body alone is sufficient to produce ecstatic transcendence in the presence of a ritual that can shape what occurs while the nervous system is stimulated. Ritual aligns the body like a key in a lock, opening the door to perception through alternative "organs" that let us see, hear, smell, and feel waves of energy that we have learned to screen out. Ritual body postures are just that: rituals performed by the body as it holds a stance in a very specific way.

To use a sacred posture, first smudge, then call the spirits as described above. After fifty belly breaths to slow down the mind, I will ask everyone to assume the posture we chose for that session. In the following chapters, twenty postures are described in detail; thirty-nine different postures are also detailed in *Ecstatic Body Postures: An Alternate Reality Workbook*.[2] Information is included about where the artwork showing the sacred posture has been found, how to assume the posture, and how it is best used given the experiences of participants in the research phase of learning about this pose. Another resource for postures is Nana Nauwald and Felicitas's book *Ecstatic Trance: The Workbook*.[3]

For many years we have been investigating what happens to us during this expanded state of consciousness that we call ecstatic trance. Brain wave research has allowed us to see that during the ecstatic trance state something is always happening in the right hemisphere of the brain, even when the left hemisphere does not register it. Using the electroencephalogram technique of spectral analysis, physicist and psychologist Gunter Haffelder of the Insitut fur Kommunication und Gehirnforschung (Institute for Communication and Brain Research) in Stuttgart measured the brain frequencies in subjects participating in an ecstatic trance session using sacred body postures. Looking at the graph, it is quite clear where the brain begins to identify the beginning of the rattling. In the case of one subject, there is the characteristic ridge of new activity in the chronospectrogram on the right side without the corresponding ridge on the left side. That individual would most likely not have identified any internal changes from an objective, observational perspective, but the monitor shows that the presence of the rattle sound stirred emotional responses. From many sources, including the writings of Sigmund Freud to Malcolm Gladwell's recent book *Blink,* we have learned that there is more going on in our psyches than we consciously know. Apparently that is true in the ecstatic trance state as well.[4]

Felicitas wrote that "ritual is the means of communication for (the Spirits), as important as speech is for us."[5] We know when a ritual is correct because something happens. What is it that happens experientially in this aroused ecstatic state? It is essentially a three-stage process. First we make the shift of consciousness and open to it. We surrender our stubborn attachment to the world as defined by ego and allow ourselves to be taken by the sound and our bodies' response into other dimensions of reality. It has been my experience in working with hundreds and hundreds of people in both workshops and counseling settings that we are never taken further than we can go while still maintaining the conscious awareness of the observing ego. The point is not to lose consciousness but to expand it. It is also true that in the expansive movement we never explode but always retain the boundary of self-aware existence within the personal self. Guided by the spirits whose home we visit, we are then trained by them to continually expand and receive, expand and receive, moving through the first stages of seeing colors and forms, then of perceiving animals and landscapes and perhaps other spirit forms, and finally of stepping into the dimension of ecstasy in which we are at one with these multiple worlds.

At the end of an ecstatic session it is time to reduce the expansiveness, to pull back slowly and carefully. In ecstatic trance workshops and groups, we suggest that when the rattling or drumming stops at the conclusion of a fifteen-minute session, people remind themselves to change their positions, to move out of the ritual posture, and to breathe slowly for a few minutes, reviewing the experiences of trance, keeping their eyes closed. Then it is best to document what has occurred during the trance session, through writing about the experience, sharing it verbally in a group session, or perhaps in some form of music or visual art. The dominant rational functioning of the left brain can override the subtler trance state, no matter how powerful it was in the moment, just as we often experience when we try to remember our dreams the next day. To retain the memory of the experience, it is important to set down memory tracks in the left brain as well.

We know that the intention with which we begin the ritual as well as the physical setting and our psychological orientation all determine to some degree what we experience during an ecstatic trance state. We see through the filter of what is meaningful to us at the time. The use of the rhythmic sound, like drumming or shaking a gourd rattle, affects the nervous system, suppressing the activity of the logical left hemisphere of the frontal cortex. We become more open to the dreamlike perception of the right hemisphere, activating the capacity to see and experience inwardly. Because we do this every night during the dream state, our experience of dreams becomes the guide to help us understand what happens during trance. Sometimes I use dreamwork techniques to help people understand the content of their trance experiences, although this is only one tool.

The surrounding physical landscape or geographic field where the ritual takes place can also affect what is experienced during a trance. Spirits traditionally known to inhabit specific locations are more available when we are in those places. Rituals activate the sacred energies of those places. Once, during a workshop on the beach in Mexico, a woman described what we were doing as "acupuncture on the body of the Earth." Stimulating our bodies with the sound of the rattle seemed to have a resonant effect on the earth beneath us. In response, we were more attuned to the presence of the Maya pantheon: the goddess Ixchel and the sacred ceiba tree, the local Tree of the World, were more present and meaningful to us there.

Finally, our worldviews affect the form visions may take. "The mind," writes neuro-psychiatrist Daniel Siegel, "emerges at the interface of interpersonal experience and the structure and function of the brain. . . . Interpersonal experiences directly influence how we mentally construct reality."[6] The people in our lives teach us how to interpret the billions of sensory messages we receive every day, and our minds create a central integrative function to make sense of it all within the personal narrative we call "self." Scientists can record the electromagnetic field that surrounds our physical bodies that some people can see as an aura and identify its varying vibrations as color. Other people do not perceive the field visually but can intuitively read the emotional energy when they enter a room or interact with an individual. Research into the functioning of mirror neurons in every human being informs us that every single one of us is reading this energy, all the time, but some of us are aware of it and others are not.

Similarly, we learn to interpret our experiences in nonordinary reality according to past experiences in ordinary life, the influence of the people around us, our physical environment, and our belief structures.

Some years ago, a couple of psychologists at the Saybrook Institute published a report on research they had conducted concerning the use of ecstatic trance postures.[7] The focus was on the Bear Spirit Posture, a pose so commonly seen around the northern hemisphere that it may have been an ideogram conveying a greeting on the order of "good health to you." The findings in the Saybrook research indicated that only a few of the subjects had actually seen a bear in their visions, and so they concluded that the method was unreliable for evoking the correct experience. Felicitas was terribly annoyed with their article. In our personal discussion and later in an unpublished paper (she submitted it for publication but it was rejected by the journal that had printed the original research report), she likened the use of an ecstatic trance posture to the use of computer software. The posture simply activates the body's capacity to function in certain ways but does not dictate the details of the visionary experience, any more than every person using Excel would come up with a similar spreadsheet.

This is a body-based practice that is difficult to thoroughly describe until you have had your own experience of it. Think about your perceived experience of a good massage or a tai chi session. The external conditions and maneuvers can be explained, but that is not sufficient to fully convey what has happened. I invite you to experience ecstatic trance.

PART 2

Experiencing Ecstatic Trance through Ritual Body Postures

In organizing our collection of ecstatic trance postures, Felicitas labeled certain categories simply to help create order. Although it is true that some postures tend to lead to initiatory experiences whereas others evoke experiences of changing form or hearing specific guidance, undertaking ecstatic trance almost always leads to some form of healing. And guidance is available through *every* trance experience. Therefore, by categorizing these postures into sections that differentiate their uses, I do not mean to limit their function, but rather to describe tendencies or better uses of one posture as compared to another. Keep in mind that in many small societies, shamans may have only used one or two postures in their rituals. That said, here are some indications of the organizing principles and definitions that would lead us to include a posture in one or another of the categories.

Healing is the inexorable draw to wholeness. All diseases—including cruelty, poverty, ignorance, self-pity, and deceit, as well as afflictions of the body—are the result of imbalance. In healing postures, we move toward greater wholeness. It is an elusive state, full of bliss at reuniting with the intelligent love out of which we emerge, while still experiencing the "I" that knows itself as separate, the observing one. Nonetheless, healing trances bring us relief and release, or the capacity for directing healing energy to others.

Divination trances take us into closer alignment with truth. We can see more clearly, sometimes beyond the constrictions of time, sometimes seeing what is perpetually in front of us but out of focus. The purpose is to discover what is hidden. For beginners we suggest asking simple questions, perhaps seeking a simple "yes" or "no" answer, then progressing to asking for clarification. The more precise the question, the more meaningful the response will be.

Metamorphosis gives us the opportunity to practice being identified with a different form other than the one we are accustomed to inhabiting. If I can be both Belinda and a bear, then I can begin to comprehend that mysterious truth that everything is a manifestation of one continuous field of energy. If I can be Bear and Spider and Swallow, then I can expand my experience of this world, to see and sense through different eyes. My usual perception is limited, whether by the spectra of light and sound my organs are capable of registering or by my limited focus of attention based on the needs of this material organism I inhabit. If I am experiencing myself in trance as a jaguar hunting in the night jungle, I see and smell and feel my body very differently than as a human woman in my house, with my refrigerator for food and my computer for documenting my mental world. When I practice metamorphosis, I can begin to comprehend nondual reality: everything is the same intelligent love of which I, too, am made. And if that is true, I am increasingly less confined by the boundaries imposed by my individual self.

While divination frees me of the strictures of time, spirit journeys open the boundaries of space. All of these experiences are ones of transcendence, going beyond the previously known limits. With spirit journeys, I can travel: around the globe in the Middle World through the spirit counterpart of Earth; into the depths of the Lower World, home of the animal spirits and the spirits of the dead; and finally to the Sky World, home of thunder and clouds, but also sun, moon, and Milky Way, and the vast beyond we call the universe. Soul retrieval is a form of spirit journey undertaken to find and reclaim what has been lost, in this case parts of oneself, one's soul.

Initiation brings us back to face the inevitable application of this expansion of awareness, to the personal experience of death. The transcendence of death is not accomplished through denial and rising above pain, but through engaging and embodying the cycles of life: the bud, the blossoming, the fruit, and then decay and barrenness, and finally rebirth in the new spring.

Each posture is shown as a drawing of one artifact and is accompanied by a description of how to physically assume the pose. We always try to stay as true to the example as possible. The only departure from this principle is in advising that you close your eyes no matter whether the eyes on the artifact are open or closed. Since we have not been trained for many years to perceive the inner worlds while also taking in the stimulation of ordinary sight, we usually cannot stay focused on the trance experience while our eyes are open. Therefore, every description includes asking you to close your eyes. The sacred postures in this collection come from all over the world and represent a wide variety of historical periods. Some were identified in only one location and others show up on several continents. Our research on the presence of this phenomenon of ritual postures used among native societies is ongoing, and Felicitas's files, which I inherited after her death, are bulging with postures that have not yet been investigated. It is an exciting field of exploration that continues with data made available through professional members of the Cuyamungue Institute, at www.cuyamungueinstitute .com, and our European counterpart, at www.cuyamungue-institut.de.

5

Healing Postures

Healing is not the same as curing. In various non-Western worldviews, physical and psychological diseases are defined in terms of imbalance throughout the whole body as well as imbalance between the individual and the society and general environment in which she lives. Curing implies making the disease go away, whereas healing is an ongoing process of attending to the imbalances as they arise. Symptoms, in this context, are indicators of what might be necessary to find or restore wholeness to the individual. Healing is holistic, requiring that attention be given to every aspect of the body, the person, and the family system, as well as the physical and social environment.

A friend of mine, a woman I have known for years, was recently diagnosed with breast cancer. In talking through the shock of this discovery, we found ourselves identifying so many contributing factors: her family history of cancer, the known toxins in the materials she handles daily as part of her work, the stress on her adrenal system as she gave of herself to her family and her work, the death of her brother, her need for self-love that was too often countered by a strict inner critic, her capacity for denial, and her many fears, including a deep distrust of traditional medicine. Clearly healing would require addressing all of these and probably more that we had not uncovered.

When we asked the Bear Spirit, the old healer Grandfather Bear, for help with her healing, both of our ecstatic trance experiences using this posture highlighted her need to know, without any doubt at all, that she could lean on Bear for support, at any hour of any day, and that the support would always be there in abundance, solid and sure.[1] No one human being could provide that. At the same time, she recognized that she needed to release some of the fear that was bound up inside her, draining her energy and straining her immune system.

My own experience during the same ecstatic trance session with the Bear Spirit Posture led me to set up a healing ritual immediately after the trance. I asked her husband, who had willingly participated so that the family system could be represented, to stand close behind her to extend the sensation of having Bear standing behind her and embracing her energetically so as to give her the feeling of support she needed. When I was healing from a divorce twenty years ago, Grandfather Bear would stand behind me in trance sessions and open his fur coat so that I could step inside him and have him draw out painful and harmful emotional energy. I knew how comforting his presence could be and how valuable it was to physically feel that solid support. In the ritual I smudged my friend's breast area, including her lungs and back, with thick sage smoke, energetically detoxifying the area, and finally directed her body's energy flow using a method similar to Dolores Krieger's Therapeutic Touch. When I asked her to pay close attention to what she was sensing in her body, she said she was focused on a ball of fear in her belly and all over her body. We agreed that she could let the fear drain out of her feet and into the towel lying under her feet. Also in my trance experience I had seen the Tlazolteotl Posture appear and invited her to do ecstatic trance once a week using this posture to help in releasing impurities from her body. When I told this to her, she smiled and reminded me that she had created a mandala dedicated to Tlazolteotl in a workshop we had both participated in the previous year. Everything was working together to give her the resources she needed to face healing from breast cancer.

My friend will continue to consult with her naturopathic physician as well as the breast cancer specialist at the hospital. However, now she has additional experiences to support her healing, both through her own ecstatic trance sessions as well as from the ritual we enacted based on our collective trances. I always encourage people to find some way to bridge their everyday consciousness with the nonordinary reality they know through trances. Indigenous healers are masters of creating these ceremonies. The purpose, in part, is to communicate with and satisfy the concrete mode of brain functioning that was our primary way of interacting with the world when we were young children.

In the ancient Huna tradition of Hawaii it is taught that every person has three aspects to the total self. The middle self is the personality that perceives and operates within the world of everyday activity. In addition to this ego that we normally know ourselves to be, there is also a high self, or Aumakua, the eternally united male and female aspects that function as spiritual parents until we are able to fully realize that level of consciousness. The third aspect is the lower self or the child self, called the Unihipili. This child self still functions the way we did as young children, needing to actually touch and see and hear rather than discuss abstract concepts. The most important thing about the Unihipili is that it is the contact point with the Aumakua, the high self. In this teaching, ordinary consciousness cannot make direct contact with the spiritually realized self, but must go through the Unihipili. Didn't Jesus say, "Except you be converted, and become as little children, you shall not enter into the kingdom of heaven"?

Psychoneuroimmunology is based on a body of research that reveals how psychologically healthy experiences maintain the immune system through the impact on the nervous system. This fascinating new field of medicine provides scientific support for what occurs during ecstatic trance states. With healing postures, we approach the experience with an intention—either for ourselves or others—to find what is needed for wholeness and restoring balance. The stories that accompany the descriptions of each of these postures suggest ways various people have understood healing and used their experiences to incorporate healing into their bodies and their lives. When in doubt about what to do, ask during your own trance experience: What can I do in my daily life or ordinary consciousness to further promote this healing?

The African Healer Posture is especially useful in supporting personal transformation into becoming a healer, with physical changes occurring that seem to empower the capacity to direct healing energy to others. In the Couple from Jenne Posture, there is both a male and a female version working in concert so that the person in the male posture donates energy to the person in the female pose. Their physical proximity is quite intimate so that partners need to be very comfortable with each other. With that closeness, the two are capable of powerful healing for themselves as well as directing healing toward others. Healing and divination are combined in the Lady of Atotonilco Posture. You can ask questions for personal advice that will support healing as well as ask for help in what will be useful in supporting the healing of others. In addition, actual healing energy is usually drawn into the body during the trance experience, often with the help of powerful spirit allies. With the Standing Woman of Jalisco Posture, participants make an "ahhh" sound in conjunction with the physical posture. People report accessing Earth energies for physical, emotional, and environmental healing that may include the balancing of primal energies, such as the yin/yang.

The African Healer Posture

The spirits are sly. One day you learn about ecstatic trance and already your body is waking up to memories that you have been carrying around for thousands of years. Your eyes start pulling you to images that previously would have warranted only a glance. You notice gestures and stances. In a subtle and demanding way, your body is looking for postures, hidden doorways that lead to that delicious ecstasy that comes from reunion with the spirits.

Diana came to a workshop in Montana and with a group of twenty other women crawled into an ancient cave to meet the Bear Spirit through the ritual we call ecstatic trance. In the darkness of those circuitous caverns, human remains from the Ice Age have been discovered, along with skulls of cave bears placed in what appear to be ritual sites. As we positioned ourselves in the Bear Spirit Posture and the rattling began, Diana was embraced and devoured by the Bear Spirit. Bear ate her, she said, so she could live inside him. It was the beginning of a series of profound and shattering experiences that lasted several years as Bear, known as the Great Healer around the world, initiated her as a healer.

Many months after her experience in the cave, Diana was strolling along the streets of a Colorado town, window-shopping. Her eyes were attracted to a dark wooden carving that stood on a small table in a curio shop. It was a statue of a bare-breasted woman dressed in a raffia skirt, with tattoos on her belly and breasts. The exaggerated angle of her arms and hands, and the deep crouch of her legs, suggested that it was a ritual posture. The shop owner knew only that the carving came from Africa.

Diana took photographs and called me to tell me about her discovery. I asked questions and she returned to the store to find out more about the statue. However, when she arrived, she found that not only was the statue gone, but the entire store was dismantled and empty. Intrigued by the mysterious disappearance, we began our research to discover whether indeed it was a ritual trance posture.

Diana used the posture for trance three times before writing to me about her experiences. "All three visions began identically: I am in a small African village. All of the huts are round with pointed roofs. . . . A stunning African woman emerges from the hut in front of me; she is physically beautiful, tall and powerful. . . . I have never met such a radiant being who is totally present."

The woman identified herself to Diana as a healer and danced her hands around Diana's body, relieving her of cold symptoms on the first occasion and later healing her lower back pain. In her third trance, she wrote: "I was telling her about having been eaten by Bear and still being inside Bear. I asked if she would help me understand this, what it meant. Suddenly she was standing beside me. She was brushing around my body, but not touching it, as though she were brushing and combing my fur. Bear loved it; he stood full upright and roared his appreciation. She continued the hand motions, going slowly round and round me, making soft sounds. Then we were outside and the villagers gathered to help us dance. She then indicated one of the men in the village who shared a similar relationship with a bird (as I had with Bear). He and I went into the center to dance. Then I found myself in utter blackness, approaching the only light there was—a small fire. A very powerful and skillful man sat across the fire from me. . . . There was a woman sitting just to my left. We were waiting for others like us to come. We waited patiently and with confidence that the other Bear-people would come." It was important for Diana to understand her role as a healer and to find a sense of belonging among others who had been called by Bear. Among others of her kind, she could rest quietly, waiting for their collective purpose to be revealed, content in knowing the challenges of the previous two years had meaning in preparing her for the healing work she was to do.

When Diana shared the African posture with her friends from a drumming circle, they also saw a woman who moved among them and danced close to them. Other groups around the country experimented with the posture and among them there was consensus that the African statue was a healing posture, even though I had told no one about Diana's preliminary experiences. For example, Elsa in California was met by a "large black woman with shiny ebony skin" and accompanied her in feeding children, healing them, and helping people die.

Deborah experienced a personal healing, as "my body, the skin of it, unravels to reveal me still standing with organs naked and exposed. . . . My ovaries are shaken up and healed. I am rewrapped and unwrapped again and again. . . . Felt my whole nervous system being enlivened, shaken up." Cinda felt her body split and water pouring from her vagina, "lubricating civilization." JT entered a cavern inlaid with crystals, bright with white light, a place for healing. Everyone reported a great fiery heat that is characteristic of healing trances.

Joan described a healing liquid filling her hand. "I knew I was a healer. I could not get healing from this posture, but I could give it." Judy told, "Energy went from my vagina to the ground . . . went like a column through my legs and my body. It built up like a well filling. . . . As I filled up, a beacon came on in my head. I could see differently. This posture was about discerning physical ailments." So Joan and Judy became healers themselves rather than being visited by the healer.

About nine years later, my husband and I were visiting the Brooklyn Art Museum and I found myself standing in front of a four-foot-high statue of the African Healer. This representation, dating from the eighteenth or nineteenth century, was identified as a traditional figure from the Songye tribe in the Congo. When I saw her, I smiled in recognition. Here was the mysterious African woman, elegant and powerful, who had first introduced herself to Diana in a tiny curio shop so many years before.

Fig. 5.1. The African Healer Posture

Performing the Posture: Stand with your feet parallel and about six inches apart. Bend your knees deeply with your back erect and your head facing forward. Cup your hands and hold them palms up in front of your body. Keeping your shoulders squared, pull your elbows back so that your cupped hands rest at waist level on either side of your torso. Face straight ahead with your eyes and mouth closed.

The Couple from Jenne Posture

The Couple from Jenne Posture is our only example, so far, of a male and a female figure represented in the same statue. The Couple from Cernavoda Posture, documented in my *Ecstatic Body Postures: An Alternate Reality Workbook,* also represents a pairing of a man and a woman, but the two figures are not joined and the male and female statues have sometimes been discovered separate from each other.[2] We see a male and a female version of the Chiltan Spirits Posture, and the Albatross Posture as well, but again in separate figures that are not always located together.[3]

The figurine we call the Couple from Jenne Posture comes from the Niger River delta, located at the bottom of the hump of Africa on the Atlantic coast. This terra-cotta figurine is exhibited in the Museum of African Art within the Smithsonian museum complex in Washington, D.C. *Smithsonian Magazine* identifies it as having been excavated near the town of Jenne and dates it from approximately 1250 CE.

The Explorers' Group, a group of people who are experienced with ecstatic trance and who want to experiment with new postures, gathers at my home every month. The group has been meeting for nearly a decade, and the six people who experimented with this posture are among the core group. At the time we first researched this posture, they had been doing trance with ritual body postures consistently for many months and had developed a sense of themselves as a tribe, with a familiarity born of sharing regular journeys in the Alternate Reality. This familiarity may have colored their experiences, so it is important to note that the trances of a different group may reveal additional characteristics of the posture.

Once we agreed to work with the Couple from Jenne, the group spent a few minutes in silence to ask for guidance about how to pair up. In our group, it seemed that the men were told that the selection of partners was not up to them, and the women were in the role of making that choice. The position is one of physical closeness and the energetic connection between the two partners tends to be strong, so it is important for everyone to feel comfortable and to be at ease.

Because both the Couple from Cernavoda Posture and the Chiltan Spirits Posture are used primarily for healing, we approached the Couple from Jenne with the question: Is this a healing posture? Several healing experiences took place. Most importantly, we noticed the same relationship between men and women as in the other healing pairs, that is, the men provided support and were a conduit for energy while the women were more active in the healing function. This was clearly a reversal for the men, who all commented on how different it felt to be in a more passive mode. John said he "felt energy being channeled through me into Michele. I asked what the Trance Posture was used for and got a message back that it was for moving healing energy and that I was like a big dish antenna pulling in the energy." The blending of male and female, yin and yang, seemed very important in creating a dynamic equilibrium that allowed greater shifts to occur.

All the men expressed learning that the details of the position were important. John was told to keep his head at a fifty-two-degree angle, and Darryl exaggerated pursing his lips, discovering that his throat was thereby more exposed. Brian and Darryl both experienced energy being cycled from the base of the spine to the throat. Brian said that his energy was similar to that of a cobra but with the snake's mouth held upward so as not to intimidate his partner. Several women felt the heaviness of their partners' hands on their shoulders and believed that the degree of pressure was important. Men and women alike described swirling, spiraling, fiery, or hot energy, and everyone was very warm after the trance.

Specific healing took place both physically and psychically. Chris saw the universe being born, as well as breasts and yonic shapes that emanated fertility. She described tubes that worked on her own body, moving and clearing energy. Her partner described wanting to send healing to Chris's body and to her friends who also struggled with breast cancer, and he could feel currents of energy flowing through him. Michele, however, experienced John's hands on her shoulders "holding me back" until she realized she was holding herself back and then she felt empty and free. The pain in her neck, from "carrying the world on my shoulders," was gone by the end of the trance. Ruth encountered a menacing male figure that her male partner was able to push away. By the end of the trance, she was strong enough to shape-shift, and she and her partner became birds that flew to the stars. They were able to get information from the star people "to help people on Earth wake up and to help them in their great confusion," which is another form of healing. While both Ruth and John each became a bird of prey—Ruth became an owl and John became a hawk—neither of them believed the posture was for the specific purpose of shape-shifting.

The powerful transformational quality of their trance experiences prompted the group to consider using the Couple from Jenne Posture to undertake a healing project together in which the group intent would be focused on a single purpose during trance. This form of group healing would be an extension of the cooperation characteristic of all the healing postures that include both a female and male form, magnifying the potential benefit.

Performing the Posture: If you are the female in the couple, kneel so that you are sitting on your feet. Place your hands palms down over the tops of your knees, keeping your fingers close together. Keep your shoulders square and jut your chin forward so that your head tilts backward. If you are the male in the couple, sit behind the woman and rest your hands, palms down and fingers close together, on your partner's shoulders. Sit with your knees raised and with your feet resting soles down on the floor on either side of your partner. Since most men do not have the long torso of the male figure in the statue, you may find it helpful to sit on cushions in order to be tall

Fig. 5.2. The Couple from Jenne Posture

enough to hold the posture comfortably. Tilt your head back as far as possible, jutting your chin forward. Both partners keep their mouths closed, but with lips puckered, and eyes closed.

The Lady of Atotonilco Posture

The Lady of Atotonilco is one of a group of "sheep-faced" figurines, so named because of their characteristic dull facial expressions and large ears. The head is elongated as the result of cranial deformation, one of the myriad ways humans have modified their bodies for the sake of beauty. The terra-cotta statue stands nineteen centimeters tall and is from the state of Jalisco in western Mexico, dating from 0–250 CE.

The body of the figure is red terra-cotta painted white on the skirt, hands, and as highlights on the jewelry and facial characteristics. Interestingly, many people reported seeing red and white during their trance experiences even though the black-and-white picture of the statue, copied from a book, did not reveal these colors. Jean interpreted a field of white as "purity and integrity" while Marti was wrapped in a soft white animal skin and Denise described snowy white summits on the horizon. I rode on a milk-white horse. Jan heard a feminine voice in a red mist, and Marti saw the sand bleeding red. Dennis began his trance with "no picture, just red."

Linda experienced dancing around a central tree, like the Tree of the World, with an old serpent with red eyes entwined around it. The serpent said, "You must look over here," and she saw the sky glowing as though it had been painted with watercolors of every hue. She was instructed then to look in the opposite direction where the colors were dark, purple and black. She stayed in the treetop all night, as an initiation. In the morning, the people below brought her a skirt and blouse, the pastel colors of the sky, and a basket of beautiful feathers of every color. Valerie saw black and purple also, but when she relaxed and her head was slightly raised, lacy white cobwebs began to show through the purple. Chris saw a purple and black rod descending from the sky and into her chest where she had had surgery for breast cancer.

Once we researched this posture the Lady of Atotonilco quickly became a favorite, mediating trances that provided both healing and divination, at least in response to personal questions. Jean said, "She (the lady) was the part of me that had the answers to everything." Her trance was a series of questions and answers, and when she paused, the lady prompted her to "Hurry up, ask me another."

Healing seemed to occur both in direct physical ways and through advice, which might be a form of personal divination. Marti received kisses "inside the hurting parts and the aching stopped." When she asked about the pills she was taking, the Lady of Atotonilco told her to "use all the resources that come to you." Heather described her healing in this way: "I am sleeping in the cool warm earth wrapped in moss. The earth draws toxins out of my body into the molten center of the earth to be recycled into new energy. Butterflies flit around me, kissing my body in hundreds of places."

In several cases, the posture was also useful in directing healing energy to others. Eagle came to Bernadette and took her on its back to fly her to her mother's hospital room, where its giant wings created a wind that cleared the anesthesia from her mother's body. Following a previous surgery, her mother had suffered from depression caused by the anesthesia and Bernadette had been worried about her recovery. Mary was transported to the critical care unit of the hospital where her friend was being treated for pneumonia. She discovered that the illness was related to her friend's connection with her son, so Mary was able to ask for healing for the son as well.

Wayne went through a shamanic transformation as a healer. First, he was "laying hands" on the heads of people but was captured and put in jail. He turned into smoke, "a healing smoke," and smoked or smudged each person in our group. "I was still followed. I was put in a box to hold me. I turned into water and seeped through the cracks. I was guided into liquid gravy. I was dropped on peoples' heads."

Workshop participants used words like "gentle," "soothing," "soft," "comforting," and "peaceful" to describe their trances. Joan said at the conclusion, "It was a warm and happy time full of golden light." Nevertheless, Helga endured a ritual dismemberment as the spirits took her to Machu Picchu. "There were some beings that circled around me. They tore off one of my arms, stripped off the flesh, pulled my bones out of my body, and when my soul appeared, which was blue, they kept jumping on it. I slipped into my soul, which made me feel good. Then I saw Machu Picchu again, Eagle appeared and put me back together, and then we flew to look at the landscape."

Eagle, Bear, and Snake—all of them powerful spirits—appeared with unusual frequency along with other helpers and guides. When Richard encountered the lady "we became snakes entwined, and the snake was her helper in healing and the rattle was the rattlesnake." He commented at the end that the snake told him the posture was good for groups needing to heal psychic schisms or splits.

Performing the Posture: Sit on the floor with both legs bent to the right and tuck your left foot under your right knee. Hold the fingers of your left hand closely together and rest your left hand on your left thigh midway between your hip and your knee. Hold your arm slightly away from your body. Keep the fingers of your right hand close together and place the pads of your fingers on your lower right cheek. Close your eyes and your mouth, and face forward.

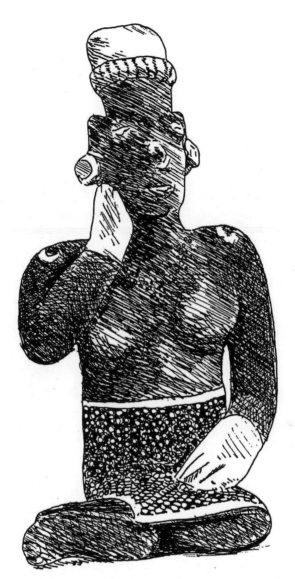

Fig. 5.3. The Lady of Atotonilco Posture

The Standing Woman of Jalisco Posture

Drawing out the purpose of a posture can be a long and slippery process. The common themes that connect individual experiences sometimes stand out immediately, whereas in other cases they elude us even after several experimental sessions. We began investigating the Standing Woman of Jalisco Posture in early 1995, and over the years, I repeatedly put the folder at the bottom of the pile, the place to which I would relegate confusing material. Finally, after seven previous sessions, we decided to try again. Paying attention to the statue's open mouth, it seemed possible that sounding a tone might be appropriate, and indeed that change helped to bring our collective experiences into clearer focus.

The group had not previously used vocalization, and we found the sound of a small rattle inadequate to encourage a robust sound. However, despite the awkwardness of our thin voices, the combined narratives after our first vocalizing session did suggest a common theme. The Standing Woman of Jalisco Posture appeared to be a healing posture with the specific function of clearing negative energy or blockages. Our second attempt—with the larger rattle and a few more people present—confirmed our earlier impressions, and later review of the reports from various groups over the past six years provided further validation.

The posture was originally brought to our attention in a German book of terra-cotta figures from Central and South America. The example shown in the drawing is a hollow statue about fifty-four centimeters tall from the Mexican state of Jalisco and dates from 200 BCE to 200 CE. The figure wears ear ornaments and red bands that are painted on her head, but nothing else.

A much older example was later discovered in a collection of Olmec figures. It is a small statue, three and a half inches tall, and carved in an elegant blue-green jade. The carving was discovered in the Mexican state of Guerrero, just south of Jalisco but far from the Olmec heartland on the Gulf of Mexico. In *The Flayed God,* a treatise on Mesoamerican history and mythology, the authors explain that Olmec contact with village cultures in Guerrero and Morelos is estimated to have occurred between 700 and 500 BCE.[1] The distinctive jade is identified as being from Guatemala. While the Olmec figure is in the same stance as the Standing Woman of Jalisco, the facial characteristics are those of a "were-jaguar" (or man-jaguar), and the clothing and ornamentation identify it as a rain spirit. The hands are not clasped around the back of the neck, as in the Standing Woman of Jalisco, but are grasping a tumpline that supports a bag carried on the back. In the bag is a stylized ear of corn. A representation of a rain spirit carrying an ear of corn strongly suggests a connection with the horticultural theme of replenishing life through an abundant harvest.

Although this does not appear to be a difficult position, many people reported feeling considerable physical pain: "I was not able to hold the posture . . . my back was in agony," "It was a most difficult posture to hold. Painful!" When sounding a tone was introduced, no one described physical pain, although one person expressed contact with a deep emotional pain. Several years earlier a woman in the California group was told during her trance, "This is an emotional healing when there are difficult things to be processed, (things that are) trapped like bubbles and causing you or anyone to suffer. Holding-in keeps you separate and causes pain, withers your energy so it cannot flow. We will help you release pain that freezes your heart and disconnects your top from your lower body. These are simple adjustments but you must surrender." Using her message as a guide, we can trace the process of healing through three stages: personal cleansing, centering and alignment, and connection among larger fields of consciousness.

There are many examples of personal cleansing throughout the experiences people reported over the years. Movement seems to rise from the feet upward: "a blue-white swirling energy rose from my feet up," "energy came up through the ground, through my chakras, out my mouth and third eye," "the floor was shaking and shook up through my body," "I felt energy come through my legs and feet and circulate," "it felt like a Roto-Rooter passed through my body." The energy can be quite strong at times; as one woman said, "It almost knocked me over." As a result of this movement of energy from below to above, something is released. Various people described it as particles of tension being swept out, impurities being released with sweat, shedding layers, or releasing and letting go of fears. In general, the experience is restorative, as something old is allowed to die and pass on.

For some, healing is accomplished in a more directed fashion. Chris said that "a blue mouth was taking energy from me," and in my own early experience a dagger opened my flesh so that grubs, which had been causing physical problems, were lifted out. A spinal twist was administered to another woman, who then felt the focus shift down into the earth while her head remained in the sky, and she could once again feel connected. These experiences introduced the next stage, which addresses personal alignment, finding one's center, and feeling grounded.

At this stage some groups reported a balancing of male and female energy, and in others the four elements were present. One woman spoke of being held secure by the vibration of the sound, allowing her to "be grounded and see a wide expanse." Although relatively few people described having animals appear in this posture, various birds, and particularly eagles and hawks, were identified as demonstrating freedom and perspective: "I could see far away," "I was told by Eagle to listen, watch, and wait."

Dancing sometimes provided the experience of being centered and grounded, feet upon the earth rhythmically creating and re-creating relationship. Bruce described being pulled toward the earth. In other words, grounding is accomplished through drawing energy down through the feet and into the earth again.

Once negative energy was cleared and the individual was realigned and centered, the focus moved to the group and even the planet. The following examples reveal a recapitulation of the clearing and connecting that occurred with individuals, but now these events were happening to the group as a whole: "A whirlwind was rushing around the circle and we could fly," "a spiral energy began rising from the center of the group, going up through the ceiling and the roof," "our chanting cleared a passage between the Lower World and the Sky World."

One woman saw in her trance that the vibratory field we were creating helped to make openings in Earth's energy field, and another member of the same group saw the group sending vibration from Earth out into the swirling galaxy, and more specifically to the constellation Orion. Someone heard our sound as "the music of the spheres" and experienced a soft wind coming back to the group from the world of the spirits, suggesting that the connection was made both with the physical galaxy and with the Alternate Reality.

One final characteristic was the appearance of a single male figure during the trance. He was described by various people as Hawaiian (with a big headdress), Egyptian, Mayan, Native American (with bird-feather wings), a "star-tetrahedron entity," and a "caped male figure who made adjustments on me." His function was never clear but his presence seemed to help direct the energies that were being released, either in an individual or in a group.

With the toning, people seemed to have fewer visual experiences. Heat was characteristic with and without sound, but some people experienced nausea without the toning. The positioning of the hands and arms affected the sound, and varying the sound seemed to have different effects. In David's trance, lower tones took him deeper into a tunnel where sad and frightened shadows lingered. Gradually they all moved up and out, carried on the voices of the rest of the group.

In summarizing the themes of perhaps sixty individual trance experiences, we must always account for the individual issues a person brings to her interaction with the spirits as well as the gestalt of the group.

Also, in giving examples that are only snapshots from the whole experience of an individual's trance, it seems to me that something of the magical blend of the personal and the archetypal is lost. To rejoin the pieces into a whole again, here is a woman's report from October 1998:

> Immediately an eagle flew in slightly above my head and put something in my mouth. For some reason I was not able to receive it. Then a thick wooden pole came through and went up and up. I saw a large rock or egg on top of the pole. (The egg) broke and yolk dripped down the pole. One branch grew out of the pole. I started to climb the pole but I was not in control of the climbing. There was a full moon at the top of the pole, and I was worried that I would climb right into it. As I got close, the silver light was lovely and huge but I knew I could not handle climbing right in. I did not. Next I was down at the bottom and my chest opened and out came a Mayan headdress and a snake who looked me right in the eye and then climbed the pole.
>
> "Then I was down low. A black jaguar paced at earth level above me, and the eagle was up at the top of the pole. I looked up and saw the early night sky and a small opening through which I saw darker sky and a few stars. I knew I was not allowed to see through the opening. I was only to know that it was there. I saw spidery lines in the sky that I knew were holding something closed.

Given the stages identified earlier, her trance might be interpreted in this way. We see at the beginning the energy coming up from the earth, but not necessarily through her as a cleansing, and she is working at establishing the relationship with that which is above her. In her words, she is "not able to receive it." It responds to her as the yolk dripping from the egg. She is attracted to what is above (the full moon) but believes she cannot handle it. A snake emerges from her chest and, after making clear eye contact, shows her how it is done. The jaguar and the eagle hold the polarities and she is permitted to see into the cosmic spiderweb that holds the worlds together.

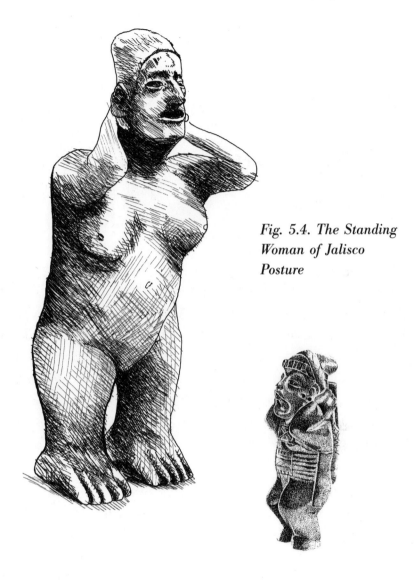

Fig. 5.4. The Standing Woman of Jalisco Posture

Performing the Posture: Stand with your feet parallel and about six inches apart, so that your hips align with your ankles. Clasp your hands loosely behind your neck, providing support but not creating tension. Extend your elbows so they are pointing forward. Keep your mouth open and your eyes closed as you face forward. As the rattling begins, sound an "ahhh," and continue making a sound throughout the trance.

6

Divination Postures

Divination is a term we in this society do not often use. More frequently people write about or describe visiting a fortune-teller, a psychic, an astrologer, or "a reader." In every case, someone is using a method that allows him or her to see and report on what cannot be easily accessed through ordinary perception, usually regarding what is hidden about a current situation or about what lies ahead in the always mysterious future. The beauty of using divinatory postures is that people can ask questions for themselves; though they may receive help in interpreting the ecstatic trance experience, they nevertheless receive the results from their own experience, rather than getting answers from someone else. Not everyone likes that level of responsibility at first, but most people come to value the certainty that arises from having an inner, visceral experience to support the new awareness or wisdom.

In workshops I often introduce divination postures by asking the group to decide on a question in which everyone is interested. It might be a question of global concern, such as "What is something we can do to manage global warming?" or "How can we collectively develop a closer relationship with the world of spirit?" Because everyone in the group has asked the same question, it is easier to translate individual experiences in light of our common focus of inquiry.

Another option is to have the group divide into pairs. Each person prepares to ask his or her own question as they enter into the divinatory trance experience. At the conclusion of the rattling session of fifteen minutes, everyone is invited to make personal notes about what occurred in the trance. Then each pair finds a private space nearby where they can share with each other their initial questions and the details of what they saw, felt, or heard in trance. Their partners listen, ask what the experience meant to the person who is sharing, ask questions for clarification as needed, and then contribute another perspective by saying, "If this were my question and this is what happened to me in the trance, this is what it would mean to me." It never fails that the partner brings insight into the meaning of the experience but in a nonintrusive way. The person whose trance is being understood can accept all, part, or none of the partner's perspective.

For many years I taught dreamwork techniques to therapists and used dreams as part of psychotherapy with my clients. The method that I distilled from the teachings of many therapists and dreamworkers was a simple one: tell the story of the dream by stating how it began, what happened, and how it ended, then identify the characters, the setting, and any key objects in the dream, and finally give the dream a title, as though it were a movie or a book. Using the technique of association taught by Carl Jung, I ask the dreamer to tell me what comes to mind when they think of each of these elements of the dream. For instance, I dreamed of a large house being built around the corner from my childhood home in which I discovered an Egyptian tomb and a group of spirit children. I titled the dream, "Discovering the Living Past." The setting overlooked a valley, which I associated with a broad perspective but not too far from home. The tomb was a little spooky, dealing with what I thought was dead but was instead still living. The dream alerted me to personality dynamics from my childhood that were still influencing me but rather than being dangerous and frightening, turned out to be a resource I could draw on for letting more creativity and play into my life.

Ecstatic trance experiences are not too different from dreams and the same method can be used to help draw out the meaning. In tribal life, in which there is a more homogenous culture, everyone in the tribe will understand the meaning of certain elements of a dream or trance, and elders whose role it is to help discern what contacts with the spirit world mean for both the individual and the community are available for advice. However, in the absence of a shared culture, it is very important to understand the social context for the dreamer or the person engaged in ecstatic trance and to honor their worldview. This respect is founded on asking questions rather than telling what a symbol or story means.

At a very practical level, divination trances can be used to assist a group in decision making. The board of directors of the Cuyamungue Institute begins every summer and winter meeting with a group divination, usually to ask what the land and the spirits need from us. A year ago the group reported, along with some specific directives, various images of water flowing. Since the Institute is located in the mountain desert of northern New Mexico where water is always an issue, we tried to focus on the water issues that might be present on our land. We had our well checked out, which led us to find out how to flush our water heater more effectively and to manage the natural chemicals in our well water. We discovered some plumbing problems that needed attention and we explored the impact of the new resort recently built in the valley below our land. (Fortunately, we found out that our well is deeply connected with a separate aquifer so that, as the well man explained, our well will likely outlive the buildings on our land. That's a relief!)

So you see that divination can cover a lot of territory, personal, global, and in the service of group efforts. As you sample the divination postures in this chapter, you will discover that different poses "specialize" in various types of questions. The Egyptian Diviner is a regal presence who gives guidance about issues related to the welfare of the group as well as individual advice for personal development and how best to serve. She is concerned about the well-being of humanity and will also answer questions of a cosmic nature. The Horned Man of Colima Posture addresses individual needs for physical and emotional healing, as well as other personal problems, and simultaneously provides direction for aligning with seasonal changes and other patterns of Earth changes. The Mayan Oracle Posture seems to connect us to a loving but unsentimental Presence who offers wise, simple, and practical advice. Finally, with the Tala Diviner Posture there is a recurrence of the theme of balancing the heat with coolness, as well as balancing right and left, and masculine and feminine energies. Indeed, the essence of divination is to become fully conscious of all polarities and to integrate them into an understanding of the whole.

The Egyptian
Diviner Posture

The Egyptian Diviner sits majestically in a small alcove in the Louvre Museum in Paris. I was so delighted to discover her that I neglected to search for and document the relevant archaeological information: where in Egypt was the statue found, from what time period, and in what temple? Her headdress identifies her as an Isis-Hathor figure, and statues of Osiris, Isis's brother-husband, reflect a similar pose but with different positioning of the arms. Another form of the posture shows her in a similar seated position with the right arm exactly the same but with an infant, probably her son, Horus, in her lap and supported by her left arm. In searching the Internet sources from the Louvre collection as well as other Egyptian sites, I have been unable to locate this specific statue of Isis to learn more about her.

Our first trance with the Egyptian Diviner posture was on October 8, 1995, on the date of a full moon. A sweet low voice of a woman, full with her own power, spoke, "We have been waiting a long time for you to return." There was a throbbing above my head and I saw the disk in her headdress as first a full moon and then as Earth radiant with blues and greens against the darkness. She instructed me to put anything in the disk and then to focus on it, and it would become like a mirror in which I could see answers. It was not hard to identify this as a divination posture.

During the trance my mind was flooded with questions. It became important to address them slowly, one at a time. As answers came pouring in, I had a strange sense of homecoming while at the same time being frightened by the power that threatened to overtake my personality. It seemed as though a force wanted to channel through me and I refused, remembering a strange and unsettling experience of spontaneous chan-neling that had occurred to me some years earlier. Nevertheless, it was so seductive to have access to answers to multitudes of questions.

I was not the only one who quickly identified this as a divining pos-ture. Everyone in our small group was drawn into asking questions in their first experience with the Egyptian Diviner, and so she was named on that first occasion. Over the years, we have learned that she is a pow-erful diviner with little patience for irrelevant or repeated questions. She is noted for responses such as these: "This isn't your problem—stay out of it," "Your heart isn't in these questions—you know you can deal with what comes up," "You already know the answer to this; ask me another question." However, when the questions are sincere and to the point, the responses are profound and meaningful. She often says, "You have to be willing, to have the courage to formulate the real questions, and to hear the answers." Chris commented, "Now that I have honored her, she is close and gives me a lot of information. She actually seems friendly now."

The Egyptian Diviner has become a guide for exploring complex practical questions about business partnerships and organizational or group matters as well as for personal development. Isis is venerated as both Queen of Heaven and the benevolent mother, so she is at home in a leadership role as well as the Divine Mother. We ask her assistance so often that it is challenging to condense my file of notes from many and varied occasions.

Among my colleagues who are business partners in a group practice that includes counseling and consulting, we have utilized ecstatic trance with the Egyptian Diviner posture to seek advice about our develop-ment as a group. An example of a fairly generic question is: What next steps should we take individually and collectively—and how can we find the energy to fund these steps?

Here is one woman's experience: "I see an image of a lion in a window. There are lions all around. The strength of our intent and power matches theirs—we are equals, and we strengthen and support each other. She tells me to learn to live with this power, to be calm, centered, grounded with it. From there comes your energy." Another member of the group saw a bingo card and was advised to "cover the whole bingo card and finally you will win. Pay attention when the numbers are called." She was amused, knowing her own tendency to withdraw and sometimes miss the lessons being offered.

In another case, I was told, "Don't assume your vision of the future is true. It may take many years for the situation to be restored. You are powerless. Face it and let go." The advice is clearly direct and to the point, and sometimes hard to swallow. In one of our best sessions, the members of the business partnership were revealed to each other as siblings, sisters who all wanted to be respected and all were very sensitive about being dominated or controlled by the others. In one person's image, there was a string tied to the ankle of each person and then tied in a web to the ankles of every other person. Everyone could still move, but it was harder to move without mutual direction and consent, or the strings would bind us tighter and the connection would become painful. This perspective on our relationships helped us understand that we could not rely on standard business practices and that we would need to apply principles we knew from family therapy to address the issues that arose among us.

In a different context, Janice shared a series of trance experiences in which the Egyptian Diviner gave her personal advice: "I heard a voice speaking as if in a ritual and I was lying on a slab of marble, cool and polished. It was almost like an operating room. I heard talking but was not totally present to understand (what was being said). I saw someone lift a shriveled heart from my chest and I heard the person say, 'This is what happens.' The heart had thorns around it that punctured others and itself. The thorns are being clipped away and (what felt like little ants) are weaving a web around my heart that allows love out as well as in. It is hard to receive love. I cried because I knew I had been mean to people who had shown love to me. What do I do?" A few weeks later, she returned to the Egyptian Diviner and the experience was a continuation on the same theme. "I couldn't remember the questions I had asked but I realized I had avoided the one I needed to ask about shame and how to get through my pain. The word 'pride' came out, loud and clear, that I was too proud to show my weaknesses. I felt Spider spinning a web over my face. The web connected us all and formed a big web that functioned as a safety net. . . . I learned the importance of group support and love for each other. This was stressed. I looked down and saw a heart crying, and heard the words, 'Listen to your heart.'" And later, "A gentle waterfall washes my body. Polar Bear holds me. My dog sits on my foot. The rattles sing." Here we witness the Divine Mother teaching her daughter, Janice, confronting her with her defenses but providing her with support from a deep sense of connection with the trance group as well as from animals in the Alternate Reality and this world as well.

The Egyptian Diviner's powerful presence requires us to be prepared to match her power with our own and not be overwhelmed. Once she advised us that we cannot be tired from a busy day and expect to come to her and experience the depths she has to offer. One evening in our trance group, a member saw herself seated in the statue looking at herself. She reported, "Her gaze stripped me naked. 'You try too hard,' she told me and I saw myself wanting to be liked and to please other people. Once I got grounded in the trance, a hose of energy burned away everything extraneous in a purifying blaze. I could see that my mentor has a stubborn willfulness that I had mistaken for an anchor in my life. It was time to get free."

In this posture, we have asked cosmic questions, too. Joan shared, "My questions kept repeating itself: What is humanity's purpose in relation to the cosmos at this particular time? The answer was in my body and was related to breathing: expansion, contraction, dissolution, finding and being one with the pulse and rhythm of the universe, with all the energies that are alive and pulsating in many forms." In her fifteen-minute experience, she witnessed what the ancient spiritual texts have described for thousands of years. The beauty of the trance is that she knew the answer in her own body and through her own breath and pulse, so that the cosmic truth and her personal essence were indeed one.

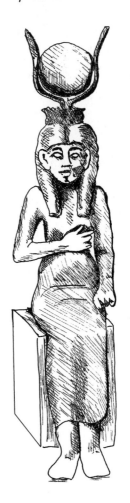

*Fig. 6.1. The Egyptian
Diviner Posture*

Performing the Posture: Sit on a chair that is tall enough so that your torso is at a ninety-degree angle from your thighs and your thighs are at a ninety-degree angle from your lower legs. Sit with your feet parallel and with your feet and knees about four inches apart. Make a fist with your left hand but keep your thumb rigid and pointing forward. With your left elbow rigid, place your thumb against your left thigh about midway between your hip and your knee. Bend your right arm at the elbow and bring your right hand to your left breast. Place the thumb of your right hand on the nipple of your left breast, point your index finger, and curl the other three fingers around your body. Sit very erect, and with mouth and eyes closed, face forward.

The Horned Man from Colima Posture

Among a wonderful collection of terra-cotta figures from Mexico is one we have called the Horned Man from Colima. Colima is a Mexican state on the Pacific Ocean that is at approximately the same latitude as Mexico City, in an area well known for pottery figurines that we have identified as being ecstatic wisdom postures. This hollow clay statue is twenty-six centimeters in height and is attributed to the period between 300 BCE to 300 CE. It is in a category designated as "thinkers," or what we know as divination postures, identified by the horn protruding from his forehead. He also appears to be wearing a cap and a pierced ear ornament.

In the description of this statue, the text points out that the fingers are unusually well defined, especially for Colima art in which they are typically carved and not distinctly molded as they are in this figure. There are only four fingers on each hand.

When we first researched this posture, we were a small group and had begun the evening talking about chakras, stimulated by questions posed by one of our members who had just returned from an intensive two-week training program for healers. As a result, we did not do our customary check-in and statement of each person's intentions for the trance. (Since we do not know in advance the specialized use of new postures, the usual practice in the Explorers' Group is to ask each person to clarify a focus so that the trance can be more clearly understood in the context of our individual issues.) We were unclear about the exact positioning of the body in this pose, since none of us could achieve the significantly curved back that allowed the Horned Man to rest his chin on his arms. However, in asking about the details of the position, Bruce was told that what is most important is the intent, which had been neglected. Then Dominic was told directly that the trance was to be used for asking questions, about "going into the darkness and extending into the future," confirming that this is a divining posture.

I had just made a difficult decision that would significantly affect the direction my work was taking. In trance, I went to the Pacific Northwest and was among tall pines and a flowing river with snowy mountains in the background. It was a bright and pristine place, in sharp contrast to the muddied feelings surrounding my work life. Salmon Boy called to me and I understood my purpose as being food for my people, an answer to my unspoken question about my work. Bruce was told this was a posture for divining weather patterns, when to plant and at what depth, how to know the amount of rain to expect so that irrigation or drainage systems could be put into place. He understood that his deeper questions were related to seasonal alignment, being part of the rhythms of Earth in a functional and healing way. Chris had also been told this was a divination posture but that it was a man's posture and not for her. A horn was trying to emerge from her forehead to help her look into a tunnel or look into the future.

The second time we used the Horned Man of Colima posture, we asked personal questions as well as a group question about a woman who had been ill for six years with symptoms no one could diagnose. Most of the group received messages about the coming winter solstice, which we had been discussing prior to the trance. Although no one asked the question, we were given information about the twelve-year cycle that would begin that year (2000) and culminate in December 2012, the beginning of a new age according to the Mayan calendar. Each person in the group also received specific information about the woman who was ill. The problem, it seemed, was related to past life experiences, and instructions were given for her to redirect her emotional energy to clear her anger and thereby relieve the twisted intestines that were the source of her undiagnosed pain.

Several group members also received detailed instructions for themselves. Michele was directed to use journal writing to face what she was afraid to know and was given guidance for dealing with seasonal affective disorder and advice about her chiropractic treatments.

Similarly, Chris got practical advice about juicing and fasting as well as clearing her emotional channels and specifically about letting go of an old friendship. Hawks, owls, and other flying creatures were prevalent. They provide a larger perspective as well as an ability to target a prey, or an issue, and dive precisely for that point.

In these two experiences, then, the Horned Man of Colima was a diviner who responded in great detail to questions about individual needs for physical and emotional healing, as well as other personal problems, and simultaneously provided direction for aligning with seasonal changes and other patterns of Earth changes.

Fig. 6.2. The Horned Man of Colima Posture

Performing the Posture: Sit on the floor with your feet about a foot apart and with your knees raised. Lean forward so that your elbows can rest on the tops of your knees. Cross your lower arms so that your right hand rests on your left knee with the left elbow on top of it. Place your lower left arm on top of your lower right arm and cup your left hand over your right elbow. Spread the fingers of both hands. Arch your back, extend your neck, and rest your chin on your left arm. Rest your upper teeth on your lower lip and close your eyes.

The Mayan Oracle Posture

The figure we call the Mayan Oracle Posture is a pottery statue 22.4 centimeters tall from the collection of the Art Museum at Princeton University. She wears the traditional *pik* that covers her body from below her breasts to her ankles. Over the pik, Mayan women wear an overblouse called a *k'ub*, sometimes made of a fine net cloth that is transparent. Remnants of blue paint cover the statue from above her breasts to midthigh, suggesting that she may have been created wearing a k'ub. Even today in the backcountry of Chiapas, Mexico, and in Guatemala it is not uncommon for women to dress only in a pik with their breasts exposed during the humid heat of the day, as bare breasts were not and are not considered immodest among the Maya. The figure in the statue also wears bracelets on each arm, a single-strand necklace with a large pendant, and ear ornaments. Such jewelry is commonly recorded in these pottery figurines of women and the real adornments worn by Mayan women were probably made of jade and sometimes of shell. Her hooded eyes and the unusual position of her left hand signaled the possibility that the statue represented a ritual posture.

In our first experience using this as a ritual posture, I went into the trance session preoccupied with concerns about a research project I had just begun. It was difficult at first to shift into the ecstatic state of consciousness as I belabored the difficult issue of how I was to be paid. Then I saw the Oracle beckon me closer with her upraised hand and she whispered in my ear a simple message that resolved the problem. She continued to give me practical advice about my work in what I described as a rich, soft, velvet voice. I knew then that this was a ritual posture useful for divination.

In the same session, Claire's vision for the Mayan Oracle was more elaborate. She was the figure "seated on a high point at the center of the city with black panthers kept nearby in a circle of protection." The Oracle held a mirror in her left hand, "used to see the past and the future together." Claire saw that people would come to the Oracle to look into her mirror, "a direct but not always happy way to learn about yourself . . . sometimes painful." In the same trance session, Jackie also saw a mirror in the Oracle's left hand. For her the experience was one of opening, being "unclogged" and allowing a purple waterfall to flow through her pores and her body openings. Seated nearby, Kathryn heard "tons of water running," and she witnessed a lush green forest with vibrantly colored birds and butterflies. Bright colors usually indicate the intensity of the experience. Barbara called the uplifted left hand "the question hand, around which possibilities flew like butterflies." Barbara also described the Oracle sitting "by the end of the ocean. The sea is deep with possibilities." In later group sessions, others also described seeing her sitting at the ocean.

Months later, the Explorers' Group decided to utilize this posture to ask a collective question: How can we shift into closer alignment with the spirits or the divine source? In response Darryl was met by "a group of masters in energy bodies . . . they laughed: our question is what being a human is all about. They are always with me. The block is the personality's stubbornness to experience guidance in a certain way," that is, according to the old paradigms. In my trance, "I was a blazing comet flying through the night sky with a burning tail, burning off layer after layer of the stuff of mental business and accumulated tension and fatigue. The tail of the comet became the image of the orphan girl from *Les Miserables*: without the stuff my personality feels orphaned but *I* am alive." Chris was advised to get connected to the earth, and Erin remembered a vision of a native person actually eating the soil in order to show her how deeply connected he was with the earth.

The next time we gathered we used the same posture and made a different request: Please give me direction about my next step on the path (i.e., the spiritual journey). She said to me, "Just live it, don't try to find words to express it yet." Darryl became an eagle in order to "get the big picture. Don't get caught in the small stuff." Alice experienced incredible pressure in her head and pain in her neck and legs. She asked during trance what to do about the pain and tried to "breathe into it" as she was advised, but she was unable to relieve it. She had a vision of asking the group for help in awakening her legs, and perhaps that was the answer to her request. John, too, was in pain, and the Oracle said to him, "Good," because he had become too complacent in his spiritual practice. Bruce, on the other hand, was told that simply being present here on Earth is the contribution we are all here to make. Rather than always moving forward or upward, "sometimes the best thing can be to step aside or fall down." Michele, a new mother, was told that it was best to lead by example, especially with children who imitate what they see. "Don't focus on weaknesses, but strengthen yourself" was the advice she was given.

In general, then, the advice of the Mayan Oracle tended to be simple and practical, to change perspective and stop trying to live by personal will, but instead to let go. She was never warm in her communication but always supportive, loving without being sentimental, revealing the wisdom that we needed to continue the journey.

Performing the Posture: Sit on the floor with your legs crossed, so that your right leg is crossed in front of your left leg. Tuck your left foot under your right thigh so that your left foot extends beyond the right side of your body. With your right arm held loosely by your side, allow your right hand to rest on the outside of your right knee. Hold your fingers together but extend your thumb away from the hand. Cup your left hand to form a "C" and turn your hand so that your palm faces your body. Bend your arm at the elbow and hold it away from your body, extending your wrist so your hand naturally tilts forward. Facing forward, keep your eyes closed and hold your mouth with your lips slightly parted.

Fig. 6.3. The Mayan Oracle Posture

The Tala Diviner Posture

There is something about the intense focus on the face of this figure that suggested it might be a divining posture. The eyes are wide open and there is a hole pierced into the ear, perhaps indicating seeing and hearing. The mouth is open too, but with the right hand held in front of it. Formed from terra-cotta, this hollow little man sits thirty-three centimeters tall. He was found in the state of Jalisco in west-central Mexico, and he is identified as being from the Tala culture, from about 200 BCE. The date places this figure a few centuries later than the Early Formative Period in western Mexico, usually identified as 1500–500 BCE. The religion in the villages in this part of Mexico at that time was a classic cult of the dead.

Back in 1994, the small group of us who were researching its use for the first time began with a question: What, as a group, do we need to do or learn together? I found myself sitting on dry, hot, barren earth that was scorched and blistered by the sun. I could see a few pine trees atop the next hill and a wind was blowing from the left. That was all.

Then a voice spoke, saying, "Wait."

For what, I wondered. "To see a vision."

"Will you show it to me?" I asked. And the voice said, "Yes."

For a long time I was just sitting, then I heard a drum, like a calm but persistent knocking on a wooden door. The drum grew louder, sounding throughout an African valley, summoning people to gather for a journey. Then all was quiet. I felt an odd duality in my body: on the left, I was cool and dark, even chilled, while on my right side, I was hot and dry, and there was brightness throughout that side of me. I wrote in my journal, "I knew I was in touch with the deepest truth of the trance, being present with both."

How could I apply this experience to our question? Waiting seemed to be a significant part of the answer. It was unclear why the earth was so hot and dry, although my later sensation of having the right side of my body feeling hot and dry gave me a clue. I associate the right side with my masculine energy. When I think of the general condition of the world, and particularly of Western societies, I think of us as being overly masculine. It was significant, then, that there was a wind blowing from the left, the feminine side. Felicitas's spirit friend always presented himself first as the wind, so I took it as a reminder that he and the spirits would make their presence known through us, as quietly but pervasively as the wind. We need only maintain the drumbeat to let the people know where we are.

In another group experience with the Tala Diviner, we asked directly, "Is this a divining posture?" The first answer I heard was, "Be patient." Then I was reminded that if I want to know if it is a divining posture, I should ask a question. I posed a question about an idea my friend Elizabeth and I had for a travel business, and I heard that it was not a business and we should not travel for money. Our purpose was to nurture spirit and the Earth. Others in that group also asked questions about personal health and their work. The common themes were the appearance of Owl and Turtle, whom we associate with deep wisdom, and the presence of a desert setting, much like the hot, dry environment of my earlier trance.

Because of the consistent imagery and sensations of going down, we thought this might be a trip to the Lower World. However, when Olga and I asked the Tala Diviner whether this was a journey or instructions for how to deepen, Olga was drawn underground, where she was told to "stay grounded and hold strong." Clear answers were given to our questions and neither of us experienced a journey.

A few years later, I thought I would ask for healing, but, "there was no sense of healing energy. Seemed more prophetic." Again, I was counseled to wait, that "something is not in place yet. It will be." At first, I thought this was my issue alone, but John commented on how long and drawn-out the experience was and that he felt as though he was being tested in his ability to endure the odd pain in his body from this simple posture. At another time, Diana asked if she could ask questions and was told, "Yes, you can ask and you can wait and wait and wait." We have concluded that this is a divining posture for diving deeply into what we want to understand. Once I even was given an old-fashioned diving helmet to assist me in diving deeply, and I have twice been advised to watch the movie *The Abyss*, in which beings of great wisdom were discovered in the depths of the ocean. Ruth traveled down a spiral staircase into a surreal world. Erin was told that every place has its hidden treasures and a tourist will never find them. In other words, we can find answers by being patient, deepening, and opening to unknown possibilities.

The heat of my first experience with the Tala Diviner has recurred repeatedly. Once my face was so hot that I thought I was sunburned from sitting outside earlier that day, and later in the trance I burst into flames. Tonia also was consumed by fire, and her smoke was pulled into the sun, clearing her with subtle heat at deeper levels. Judy said that she was going down and became very, very hot. Jeanne, too, felt a lot of heat, and Celeste witnessed smoke coming from the ground in a double helix.

Finally, there is a recurrence of the theme of balancing the heat with coolness, as well as balancing right and left. Chris said there were two energy forms hitting together and she was intent on clearing blockages and balancing these masculine and feminine energies. Jackie reported, "A lightning bolt set the lake on fire, but we were flying like hawks and dove into the lake where we sizzled to coolness and emerged brownish and wiser. We were treated as leaders, maybe treated as being wiser than we felt." Perhaps her experience clarifies the Tala Diviner's purpose. As we experience the great heat of the sun, of consciousness, of Apollo and the masculine archetype, we dive deep into the cool, dark waters of the feminine, spiritual, unconscious worlds and emerge with the wisdom of the polarities in balance. This is divining at its best, going deep in order to become fully conscious and to be able to fly. It is the essence of the mystery teachings of the cult of the dead.

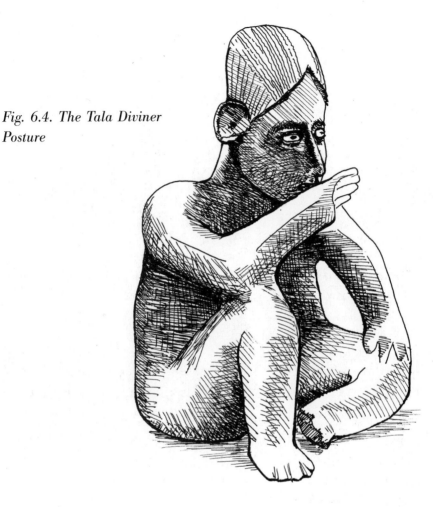

Fig. 6.4. The Tala Diviner Posture

Performing the Posture: Sit on the floor with your right knee raised and your right foot on the floor but with your heel slightly raised so that the pressure is on the ball of your foot. Rest your left leg on the floor and bend it so that your left foot is next to your right foot, with the bottom of your left foot facing your right foot. Place your right elbow on your right knee and bring the back of your right hand to your mouth so that your lips touch the back of your hand. Keep your mouth open. Keep the fingers of your right hand slightly spread. Keep the fingers of your left hand slightly spread and rest your hand lightly on your left knee. Lock your left elbow and extend your arm away from your body. Close your eyes as you face forward.

7

Metamorphosis Postures

The hunter-gatherer way of life is the oldest and most enduring of human social organization, named for the way people were able to feed themselves and thus survive. Generally the earliest men scavenged the remains of the prey of large hunting animals and then eventually learned to hunt themselves. At the same time, the women hunted small animals, rabbits, rodents, and birds, but supplemented this diet with nuts, roots, and berries found in their home environment. In this hunting-gathering life, humans were fully engaged in the intimate, and some would say magical, connection with animals. The shamans or spiritual specialists among these tribal people derived their power from sharing a deep identification with animal spirit allies, and often this was proclaimed with wearing furs, feathers, or headdresses of the animal to whose clan they belonged.

When humans took up small-scale horticulture, in which the women who had learned the ways of plants began to collect seeds and cultivate plants of their own, the intimate connection with the animals was slightly weakened. One way the spirits may have shown humans how to renew that link is through shape-shifting, or metamorphosis, in which an individual can, for a brief time, restore the capacity to

88

experience the essence of being an animal. There are stories about powerful men and women shamans who could transform themselves into the animal or bird right in front of the eyes of the community. Once Felicitas and I were talking about such a story and I asked her if she thought that transformation actually took place in the physical form. Assuring me that anything is possible, she nevertheless believed that what these powerful medicine people were able to do was to induce a trance state among the entire group and thereby cause them to witness the physical form of the animal ally in the Alternate Reality. Therefore, the metamorphosis was not a fully physical manifestation but was real nonetheless.

When we use the metamorphosis postures it is not to transform for the benefit of the group but to restore our own capacity for being aligned with the animal spirits. Our physical bodies can actually sense the development of a snout or wings, can find the grace of fluid muscles and the power of seeing in the darkness. This metamorphosis is the basis of the Masked Trance Dance workshops and rituals that are a centerpiece of our work at the Cuyamungue Institute. Through trance we invite an animal spirit to befriend us and teach us, and we in return create masks and costumes to more completely take on the animal form. The ritual dance that evolves from integrating our collective trance experiences tells a mythic story and culminates in a metamorphosis dance in which we give ourselves completely, but consciously, to expressing the essence of the animal.

The walls of my sacred space at home are filled with masks from dances in which I have participated over the past twenty-five years. In the worldview of the hunter-gatherers, all things in the material world have a spirit in the Alternate Reality, even so-called inanimate objects, meaning literally that they do not have "souls." Since the masks have spirits that have been awakened in the ritual dance, we always ask people to continue to honor the masks, and hence the animal spirits, by giving them a gift of tobacco or cornmeal. Some years ago a group of native elders directed a request to the curators of museums that housed the sacred objects of their tribes, asking them to allow them to "feed" and honor these objects that were known to contain immense power from their history of ritual use. The elders believed that some of the chaos in modern cities is the result of these powerful forces that, not being guided in a ritual way, were taken up by inexperienced youth and people who were not healthy in spirit.

One of my most beautiful metamorphosis experiences occurred in a dance when Grandmother Spider taught me to weave her web, connecting all things in this world and in the spirit world. The isolation of individualized consciousness dissolved and I experienced being at one with everything. Even in the great expanses of the Milky Way, I discovered myself always connected in the web of life. This is what the later agricultural religions—Buddhism, Hinduism, Islam, Christianity—teach as the ultimate spiritual experience, known by many terms such as *satori* and *salvation*. It is what we know as the ultimate ecstasy.

In selecting which metamorphosis posture to use, you can consider what elements of shape-shifting are highlighted by each of these poses. In the Cernunnos Posture, people became trees as well as animals, and an embodied experience of the fertility of the earth was included as part of many trances. Our experiences with the Chichén Itzá Posture have focused quite a lot on making the transformation into shape-shifting as well as leading most people into metamorphosis as birds and occasionally serpents, the animal spirits that are highlighted in the ruins of Chichén Itzá. The Fish Woman Posture is linked with the life cycle of plants, the metamorphosis from seed to sprout to plant and fruit, and with the water so necessary to support this cycle. Her evocative facial expression with her huge open mouth challenges us to surrender to a primal and passionate experience. The Olmec Medicine Man is from Mexico, like the Chichén Itzá Posture, and so it is not surprising that most people experienced becoming either birds or snakes. The Olmec culture is a very early one, however, and we find similarities between this posture and others found in Egypt from about the same era.

The Cernunnos Posture

Cernunnos is a Celtic god of fertility and is one of the most widespread of the deities or spirits from Celtic culture. His image is displayed on a cauldron found at Gundestrupp in Denmark, from the third century BCE. He wears antlers, holds a torque in one hand and a snake in the other, and is surrounded by lions, dragons, elephants, boars, wolves, birds of prey, and a boy on a dolphin. In the Celtic world, the torque represents power and authority, much like the scepter held by European royalty or the Egyptian pharaoh's crook and flail. The snake denotes wisdom and sacred knowledge of the inner worlds, while the antlers show strength and mastery. The stag was able to transverse the divide between the material and the spiritual worlds. It was the All-Mother Anu who gave birth to Cernunnos. His role was to sing the souls of the dead to the Summerland, the place of reunion and renewal, and he was the protector of the animals and the forest, making it fitting that his posture would be one used for metamorphosis or shape-shifting so that humans can learn from their animal ancestors. However, this is no ordinary shape-shifting. The power of fertility and the fecundity of the earth are included in the rituals we experienced, and some people became trees instead of animals.

At first I had trouble understanding this as a metamorphosis posture because there seemed to be so few animals. Many people in Germany as well as the United States interacted with trees, "growing high into the sky and deep into the earth." Geoff's shoulders and body were covered with vines before he actually became the World Tree, with the snake eating his roots and deer eating his leaves. He related this to a Norse myth of Yggdrasil, the World Tree that grew in the hall of Odin where animals derived nourishment from its leaves. Sometimes in spirit journeys we travel along the roots, trunk, or branches of the Tree of the World, but no one actually traveled, so that did not seem to be a spirit journey posture. Thinking that it might be a divination posture, I tried to ask questions. At the time my husband and I were exploring a move to the Southwest, and I asked about how to find the best place to live. The answer was simple and matter-of-fact: "Where you are is the place you live." Then, in the trance, I was transported to a giant tree surrounded by animals and was briefly allowed to embody the soul of the forest. In reviewing my notes of reports from several sessions, it finally became clear that this posture requires a lot of energy and for many people the full metamorphosis had not occurred so there was less evidence of actually becoming another life form as we expected from a shape-shifting posture.

As in other metamorphosis experiences, many people felt movement and spiraling energy. Olivia said it was hard for her to concentrate but she intuited that the posture was about movement and transformation. During the same session, Michele was spiraling through a yellow tunnel with kaleidoscopic designs, Chris found the crown of her head twirling, and Bruce was spinning clockwise. Barbara said a drum was spinning like a tornado, kindling a fire that allowed her to hover between the future and the past and marvel at the vastness of the world.

For everyone, there was a lot of heat and many complaints of joint pain and exhaustion at the end of the fifteen minutes of rattling, indicating how much energy was required and how painful it can be when the energy flow is blocked. Sharon was first sitting in a deep dark pool of an underground spring, the powerful, life-giving source. Her body felt dense and she learned to become less dense to allow light and electrical energy to come through her. David encountered a snake and asked what to do about the pain, and the snake said, "Change." Chris was taken on the back of a snake into the earth and became a snake and was told that the posture was for the purpose of experiencing the power of the animals. For Olga, the snake appeared as the uroboros, the snake biting its tail that was an ancient symbol for the virgin goddess who gave birth to the cosmos out of her own body and was expressly a symbol of the world before duality. John actually shape-shifted into a deer, jumping through the woods, and Bruce grew to be about ten feet tall.

The theme of fertility appeared with many people. Pam saw a maiden having sex with the Green Man who is the embodiment of the life force in the plant kingdom and is often interpreted as symbolizing rebirth and the cycle of growth each spring. Pam eventually became the maiden and with considerable blushing told us that she was having sex with a tree, concluding from her experience that all things are interdependent. The Green Man also appeared in Jennifer's vision, leading her across the stone fence of a vegetable garden, through the gate and into the deep forest, to a magical circle of trees and "the seed of my own potential." Both Jennifer and Olga saw a woman in a white dress during this session, but we do not know enough about the Celtic spirit traditions to recognize her. Bruce saw the earth and her power come together so that he could drop deeper into the trance, where he felt the cycles of the seasons in synchronization. From several sources, dried blood was mixed with the soil, an ancient practice of fertilization. In the process, the principle of reciprocity was confirmed, "something was taken and something was given," so that the cycles of life and productivity could continue.

Jewelene met a serpent who was stretched out in front of her, showing its vulnerability. Little snakes were on tree branches and were made only of batik cloth, again making them as unthreatening as possible. When the snake opened its mouth to invite her to be swallowed, and thereby enter the metamorphosis, and even provided a light in the tunnel of its throat, she saw lots of teeth (even though snakes do not have teeth) and resisted. Her experience affirms that we always have choice in our encounters with the spirit world, and that we will not be devoured or overcome against our will.

*Fig. 7.1. The
Cernunnos Posture*

Performing the Posture: Sit on the floor with your legs crossed in front of you. Bring your right calf as close to your right thigh as possible, so that your foot is resting in front of your crotch. Position your left leg so that the calf rests in front of your right leg. With your shoulders squared, hold your arms to the side of your body and bend your elbows so that your hands are at shoulder level, creating a V shape. In his right hand Cernunnos holds a torque with the front of his hand facing forward and the fingers and thumb curled around the ring. We have used small grapevine wreaths as a substitute for the metal ring. In his left hand he holds a snake with his hands just below the head and the front of the hand facing forward. The snake's head is facing the figure. While the fabric snakes we bought are not proportionately as large as the one in this posture, we did not seem to have a problem. We have not replicated the horned headdress that Cernunnos wears, although that could be easily done with sticks attached to a cap. Face forward with your eyes and mouth slightly open as though you were preparing to whistle.

The Chichén Itzá Posture

Often Felicitas and I would comment on a seemingly incongruous situation in our groups and workshops: people new to the ecstatic body postures and similar practices would unaccountably have elaborate and classic experiences that would dazzle us all and outshine those of more experienced "trancers." Or, the first time we would try an unknown posture the reports would be the most clear and definitive about the primary purpose for which the posture could be used. Felicitas used to say that the spirits favored naive subjects who came open and innocent to the encounter. We might consider that these were circumstances in which the tendency to anticipate what we should experience was diminished, thereby reducing any bias in perception during the heightened state of consciousness that we call ecstatic trance. In any case, naive perception was a boon in 1995 when I experienced the Chichén Itzá Posture for the first time. Without the context set by that initial occasion, it would have been difficult to discern that this is a posture best used for metamorphosis.

A little background about the culture of the people from Chichén Itzá helps in understanding the posture. A colleague in Europe had sent Felicitas a photocopy of the statue, labeled only "Maya figure, Chichén Itzá, Yukatan." Chichén Itzá, translated as "at the mouth of the well of the Itza people," is the most widely known and popularly visited of the Maya ruins in Central America. The central temple, known as El Castillo (The Castle), was built in homage to Quetzalcoatl, the Feathered Serpent, known in Maya as Kukulcan. On the equinox, when north and south points of Venus rise above the horizon, the light at sunrise descends on the steps of El Castillo to create a serpentlike movement, culminating at the giant serpent heads at the base of the staircases, honoring the Feathered Serpent. Excavations uncovered an older temple beneath the pyramid in which there was a red jaguar with jade spots and one of many Chac-mools, reclining figures presumed to receive the bloody offerings to Chaac, the rain god. The sacred cenote at Chichén Itzá is a deep natural pool in the limestone layer that covers the Yucatán peninsula of Mexico and was a pilgrimage site for centuries. The remains of gold, pottery, jade, and human sacrifices made to Chaac have been recovered from the cenote. The Temple of the Warriors is another prominent building, honoring warriors of the Jaguar, Eagle, and Coyote clans. Throughout the complex, crisscross designs abound, acknowledging the power of Serpent, particularly Rattlesnake.

Chichén Itzá was settled in the Late Classic period, as early as 400 CE, around the time of the fragmentation of Tikal in Guatemala, and continued until around 1300. It was really more of a Maya-Toltec city, populated in part by exiles from Tula who were expelled by Tezcatlipoca in 987 CE. The cultural diffusion created by the mixture of these two groups would have disrupted the spiritual practices of the earlier Maya and challenged their rituals for connecting with the spirit world. Metamorphosis as a means of uniting with the spirits was already less prominent among city-state dwellers. Here in Chichén Itzá, the eagle, the serpent, and the jaguar received special tribute, so their presence may have still been strong. This posture would perhaps have been the remnants of the older practice of using ritual postures to join with them, surrounded by the later Toltec and Aztec practices of offering blood sacrifice to achieve that same end. This would explain why this meta-morphosis posture does not generate the wide circles of animal spirits to which we are more accustomed in other postures like the Olmec Prince or the Tattooed Jaguar Postures.

In our first experience, only Carol and I were doing the ritual together. In my journal I wrote that I saw first an eagle that changed into an owl, then a blackbird flying with a shaman in its mouth. The images were in the design of the native people of the Northwest Pacific Coast. The shaman began to fly on his own, and then I became the shaman, flying over a flat gray-green land. The Yucatán peninsula of Mexico is very flat and covered with jungle forest, although this may have been northern tundra. I was flying to a council meeting where a robin announced the gathering of the circle of bears, eagles, and humans. As I joined the circle, "I slipped into Bear, was Bear. Just sat, feeling it."

Later in the experience I was invited to play in a mountain meadow with Morgan, my friend's dog who had died several years before. She ran ahead, pausing to look back at me as if to encourage me to follow. It was then that I realized I was still Bear, lumbering at a surprisingly fast speed. I sat with the power of Bear as the great healer, but when I tried to direct healing energy to people, I saw them embraced by plants and flowers. At the time I thought it meant that I was to learn herbal lore, but now I recognize that this is not a healing posture and does not generate extra energy to transmit, but rather requires extra energy for metamorphosis.

In the same ritual, Carol also was flying with a bird, and in the reports of people in later trance experiences, shape-shifting into birds, and especially eagles, was the predominant theme. Many people told about time slowing down, and John interacted with a crocodile who taught him about how slowly things sometimes change. For others, spinning was the means of making the transition to another form, and, for many, this was accompanied by heat. One woman described a woven God's-eye with long arms ending in black tassels, turning and stopping, turning and stopping. Judy also reported a spinning design that helped her to go deeper into trance, while Joan's spinning with a pinwheel shape eventually caused her face to peel off. Barbara felt waves of energy coming through her, and Carol was "going through a tunnel with wavy walls." For me, in a later ritual, the rattle played a polka tune that started my cells and molecules dancing and they spun out from my body, opening to wider and wider spaces.

Once the metamorphosis was accomplished, people reported either seeing birds flying or flying as a bird. "An eagle swooped down, got

Felicitas by the nape of her neck, and flew off with her," "we were birds flying or dolphins swimming; everything was fluid," "we changed into birds and went hunting for water," "I flew to the eagle aerie in the crisp, cold air of the high mountains." For some the flight was ecstatic but others were cautious: "The lightness of flying was lovely but a little frightening, like how I am uneasy when my feet don't touch the bottom when I am swimming." Along with eagles were an abundance of owls, a heron, and several reports of butterflies. One person described seeing a woman with a colorful Elizabethan collar opening her arms and becoming a human butterfly. In Western mythology, of course, butterflies are a common representation of the process of metamorphosis.

In one session, Dominic, John, and Ruth all became snakes and two of them reported having two heads. The balance of the two heads intertwining seemed to be a lesson in integration. This challenge of balance is reflected more clearly in Alice's experience. She spoke of sitting with a friend overlooking a river at sunset with butterflies flying around. The sight evoked anger at the self-centeredness of humans. "Even the best of us don't get it, the predicament we bring upon ourselves" by our egocentric behavior. Then she saw a familiar bird from earlier trance experiences and she flew with the bird, over mountains and valleys, until she returned to see herself watching the sunset. "They are both me," she said, the bird and the errant human. Metamorphosis can teach us about our self-centeredness.

The magic of metamorphosis is expressed in Anne's telling about the stalk of corn that talked to her, "inviting me into the cob, with the promise of telling me secrets." She became tiny and it was easy to enter the warm and comfortable cob, to be one with it, looking up at the stars and planets. She said that the corn did not share its secrets with her, but I think the experience itself was the secret, the treasure that we have rediscovered in our ability to shape-shift.

Fig. 7.2. The Chichén Itzá Posture

Performing the Posture: Sit cross-legged, with your left leg in front of your right leg. Place your left hand palm down over your left knee, so that your kneecap is covered. Place your right hand palm down against your stomach, so that the palm is over the center of your body just above your waist. Hold the fingers of both hands close together. The elbow of your right arm should be held out from your body. Keep your mouth slightly open and with your eyes closed face straight ahead.

The Fish Woman Posture

This evocative image of an old woman already undergoing a metamorphosis was discovered in a wonderful book, *The Olmec World,* published by the Art Museum at Princeton University to accompany an exhibit of the same name. This white clay vessel was discovered in the area of Santa Cruz, in the Mexican state of Morelos, dating from 1200 to 900 BCE. The catalog describes it in this way:

> In the form of a kneeling, skeletonized figure, this vessel is one of the most powerful and haunting images to have survived not just from the Early Formative period but from all pre-Columbian Mexico. While this vessel from the highlands shows little Olmec stylistic influence, it embodies themes at the heart of Olmec ritual. . . . Although depicted as an ancient crone with dry, sagging breasts, the swollen belly has been taken to indicate pregnancy, and the yawning mouth, the spout of the vessel, with the greatly distended lower jaw and lip as a howl of pain . . . the kneeling position, with the head thrown back, may be that of supplication. There is convincing visual evidence in Olmec art of the ritual focus on rain, and drought must have been a seasonal and constant concern. Perhaps this creature "in extremis" addresses such a desperate time and implores the heavens for the rain on which the harvest and the survival of its issue depends.[1]

Because we approach this figure experientially rather than as scholars, our bodies tell us a somewhat different story than the speculations of archaeologists. Pregnancy does not seem to be an aspect of this ritual, although water is significantly present in the various reports from people using this posture. Michele heard a waterfall nearby throughout her trance, and Jill was standing on the ledge of a waterfall from which she dove into the pool of water where she said she harvested plants and starlight, receiving sustenance from the water. Likewise, Miriam was covered by a waterfall and was drinking at the outset of her trance, and a snake moved through her head, "like cleaning a pipe." For me, a small waterfall was emptying into the pool in my mouth and washed through me, cleansing me of grief after the death of a beloved cat. For others, the statue was part of a well or fountain, with water pouring or "spitting out of me." One person saw murals of fish swimming in water and another felt herself become the Fish Woman.

When I first asked what this posture could be used for, the answer was "Change," not an uncommon response when the ritual is one for metamorphosis. During the trance, "my molecules are stirring wildly and I go through many forms: molten lead, a dandelion in seed, an alligator. My body is only one of many, many possibilities. I feel myself in a fire, burning, and it is both painful and freeing. I am sweating, writhing, burning, but she (the Fish Woman) says 'Don't hold back from all this—it is not gentle, don't be afraid of it.'" Afterward I commented that I felt very alive. Similarly, Bruce was taught a lesson in surrender. "It was a wild ride," he said, "and I am happy to have survived it." We spoke of feeling primal and of vital energy flowing through us at the conclusion of the ritual.

The actual shape-shifting experience came through John's report when he said, "Jaguar moved into my body and possessed me, then jumped out and I invited him in again. This happened several times. My head was hot and I felt my blood circulating." Christine's lips and tongue became painfully dry and her breathing was fading. She felt as though she were dying of suffocation when suddenly her body changed. "My skin was itchy and I grew fur, then I was prowling in the mountains and I was a mountain lion. I was wide open and received so much information from my nose." Meredith described becoming a lizard with a long fluttering tongue, and Carol became an otter, swimming with the fish, and then was met by a wolf, an eagle, and serpents. The gathering of other animals often occurs along with the actual shape-shifting.

The theme of transformation is linked with the life cycle of plants, the metamorphosis from seed to sprout to plant and fruit, so common in the rituals of the horticulturalists of Mesoamerica. "Snake carried me over the fields and left me where they were planting," and later, "an Olmec figure with feathers in his hat or headdress ate all my fears and transformed them into golden packets of seeds that he eliminated onto the ground." Frances participated in the fertilization of the earth as the sun shot eggs of light into her open mouth and then she shot them into the earth.

The metaphors indicating metamorphosis vary. For Barbara it was the rebuilding of a three-story structure that we recognized as herself, then the rebuilding of the ancient Mexican city of Teotihuacán. Anne went into darkness, and then saw the Jemez Mountains, the view from the Cuyamungue Institute, turned sideways, and she exclaimed that "the whole world has turned." Indeed our identity can be turned upside down when we yield to the transformation of shape-shifting and relinquish our old concepts about who we really are and the boundaries we have previously learned to construct to separate from the rest of the living world.

Performing the Posture: Kneel with your knees spread apart and sit on your feet, or preferably the floor between your feet. Hold your arms close to your body and place your hands on the outside of your thighs just above your knees. Spread your fingers. Lean forward slightly and tilt your head backward as far as it will go, resting on your lower neck. Close your eyes and open your mouth wide, with your lower lip extended forward as far as possible.

Fig. 7.3. The Fish Woman Posture

The Olmec Medicine Man Posture

At a November gathering of the Explorers' Group, John brought a photograph of a sculpture he had discovered in La Venta Museum Park in Villahermosa, the capital city of the state of Tabasco in Mexico. The park is a lush, junglelike setting for fifty Olmec sculptures unearthed in the 1930s at the site known as La Venta, a ceremonial center for the Olmec located about eighty miles west of Villahermosa. The sculptures were moved to the park where they could be preserved and made more accessible to the public. Made of stone, they are larger than life-size, and in this natural setting they convey a powerful sense of the Olmec civilization about which anthropologists and historians know so little.

The formidable figure of this priest-king, known simply as La Venta Monument 77, sits cross-legged with his hands in a position reminiscent of Egyptian statues with their fists on their knees. When I suggested to Felicitas that we call him the Olmec Priest, she proposed that Medicine Man would be more accurate to describe his position in the Olmec culture. And so he was named.

Having met each month for several years, the Explorers' Group has developed a general approach to working with new postures. Usually a few of us remember to ask a specific question during the trance to determine the primary use for the posture. Those answers help us when the group members compare their experiences. For instance, I had a specific question for the spirits about whether to make a trip and what the purpose of the trip would be. At the beginning of the rattling there was a brief verbal message, *"Go,"* and then, *"Celebrate,"* but nothing of the detailed responses characteristic of a true divination posture. John also asked a question and received only a cryptic response before going on into the full trance. One woman in our group had just ended a relationship; she needed emotional healing and that is what the spirits gave her. Janis also experienced an involved healing journey because that is what she, too, asked for. These reports reinforce our continuing experience that individuals get what they need during ecstatic trance, even when it is different from the primary focus of a particular posture.

The primary theme of the Olmec Medicine Man is metamorphosis, changing into another form, the quintessential spiritual experience for horticultural people like the Olmec. Most of the group saw or became snakes. Joan describes being "pulled forward and feeling magnetized to be prone. I'm a snake in the forest floor, gliding over moist ground smelling earthiness with my tongue." Another woman had to work up to the snake experience, admitting that it was "a little intimidating" and that she was afraid of it. To help her along, her initial contact was with a lizard whose "molecular and cellular structures were moving very rapidly so he was not solid," a beautiful description of the process of metamorphosis. Chris saw a spitting cobra, about to strike, and she knew the snake was a doorway. David became the cobra and afterward described seeing his body as part snake and part human, as is so often expressed in sculptures and other representations of metamorphosis, such as the one we see in this statue.

After a powerful experience on a personal theme, Janis was joined by birds that were "welcoming me to their realm of sky spirits." They transported her to snowy peaks and she thought she might die but was unafraid because "it was so beautiful and awesome." Others commented on thinking they might be dying but feeling unafraid. We believe the dying is part of the experience of losing one's familiar physical form and identity during metamorphosis. However, one person experienced joining the clouds and bringing rain to her people, which, unbeknownst to her, is a classic rendition of the Pueblo Indian belief about what happens to the spirit after death. She reported, "It's done (the death) but I feel no pain. Life energy leaves the body, dripping droplets of water that fall like rain on the earth and the inhabitants below. People with painted faces are dancing and looking up joyfully."

Many described a kaleidoscope of color, with red predominant, sometimes moving to orange, sometimes to purple. As happens in a metamorphosis trance, it is so difficult to raise enough energy to complete the metamorphosis that sometimes people have only a partial experience, and in this case, several people saw only black and white. Many members of the group felt twitching, tingling, or a jolt. Michele reported, "Toward the end I felt a tingling throughout my body and it felt great, and then I felt myself slowly coming back into my body."

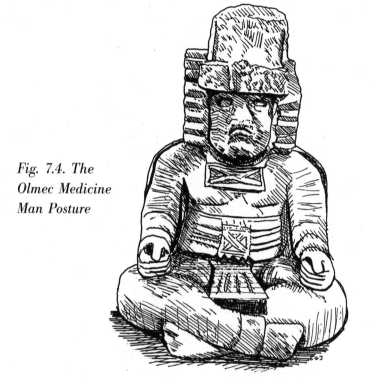

Fig. 7.4. The Olmec Medicine Man Posture

Performing the Posture: Sit cross-legged with your right leg in front of your left leg. Let your right foot rest in front of, not underneath, your left knee and allow your left foot to stick out from under your right leg or knee. Make your hands into loose fists and, extending your arms, rest your fists on your knees. Place your thumbs on top of your fists, pointing forward. Face straight ahead with your eyes closed and your mouth drooping open in the characteristic Olmec fashion. Allow your tongue to protrude from between your lips.

8

Spirit Journey Postures

Many ancient traditions tell of making journeys along a giant tree that sits in the center of the world, whether that is the ceiba tree of the Maya, or Yggdrasil, from the old Norse myths discovered in Iceland. We call this the Tree of the World, or the World Tree. Its roots burrow deep into the Lower World, home of the animal spirits and the realm where the spirits of the dead reside until they make their transition into the upper spirit world. The trunk of the Tree of the World stands as a spirit presence in the Middle World, which is our material home, and journeying along this trunk allows us to travel around to other places in this world. The branches and leaves of the Tree of the World reach high into the Sky World, or the Upper World, which extends into the stars and give us access to that realm while we are still alive in this one. So in spirit journey postures, we can go to the Lower World, the Middle World, or the Upper World, usually depending on which posture we choose.

The Bird Woman of Egypt Posture offers an experience of physical strength and power, along with the empowerment to face challenges. The spirit journey is one to the Upper World and mostly people find themselves flying. In the Upper World, when the power is great enough, their journey might extend even to the sun. By contrast, the Sleeping Lady of Malta Posture facilitates a journey to the Lower World, and people usually descend in some manner. Since the Realm of the Dead is located in the Lower World, this journey is sometimes connected with indirect experiences of death or dying, although learning about death in the cycle of birth and rebirth most often takes place through an initiation posture, found in the next chapter. The Poppy Pod Woman Posture takes us across the Middle World to witness the rich and bountiful earth, sometimes participating in the cycles of seasons. Even though it is a trip through the Middle World, the entrance still might be found by going down into a hole or into water. Finally, in the Tsimshian Shamaness Posture, we journey again into the Upper World although there seems to be a possibility of also traveling in time. Felicitas once reported going back into her personal past to retrieve memories for her autobiography, and another participant journeyed back into the distant past of humanity.

In this world of Internet and global transportation by airplanes, the miracle of the spirit journey may not seem as compelling as it was to people who never left the valley in which they were born. Yet even though I travel quite a lot with my work, it is nothing like the freedom of spirit journeys. Recently, in a trance with the Tsimshian Shamaness, I found myself riding a horse in the Caucasus Mountains with a bird flying above us, both the horse and the bird awaiting my direction: Where shall we go? What shall we find? I wanted to go to the Cuyamungue Institute, a place I love so much and yet am able to spend only a few weeks a year. It seemed as though I was on a soul-retrieval mission for the Institute and for myself, and the first thing I found was pottery shards in the snow at Tsankawe, an ancient site that is part of the Bandalier National Monument that is north of the Institute. It is a spot I love, and I found myself sinking into a deep sense of satisfaction. Suddenly in the trance a woman with long black hair and a painted face appeared, and, raising her bow, shot a rubber-tipped arrow at me. As a female warrior she was sharing her energy with me and with the

Institute. So, I mused, it is warrior energy that we have needed to give us a sense of empowerment and the passion to engage in new projects to extend our work. Enlivened, I embraced her and united with her. The raven on the shamaness's headdress transformed into a dove with an olive branch, and I was nourished with the deep peace that is the ultimate goal of the warrior.

# The Bird Woman of Egypt Posture

With arms powerfully raised, her breasts and heart exposed, this bird-headed figure dates from 4000 BCE in predynastic Egypt. She is also said to have a snake body, becoming another form of the winged serpent similar to the Cadeusus or Quetzlcoatl in later Greek and Aztec traditions, or to the winged solar disc where the kundalini rises to meet the Self for the full activation of human potential.

Consistently people using this posture report significant heat during the trance. Maintaining a pose with arms uplifted is very strenuous and requires the heart to pump harder, creating body heat. Although heat is often associated with healing, there were many reports to support the hypothesis that this was a spirit journey posture. The most obvious were experiences of flying, or of seeing or becoming birds. Ruth was soaring above a deep crevice and Ed was looking down the gullet of a long-necked goose, seeing down into the darkness. Bruce was a hundred feet above the earth. Linda found herself on "a high mountaintop that might have been a pyramid." Judy reported flying around the world with Dove as the messenger of peace. Jessica was a baby eagle "wanting to fly but I couldn't." As we will see, personal resistance kept her from being able to rise.

The physical position is one that fosters a feeling of power and the capacity to accomplish a challenging journey. One woman said, "The posture feels so good in my body: all lined up, curving, and powerful. [The Bird Woman] tells me to be empowered, don't hold back, and be at peace." Empowerment came in an immediate and practical way for Chris. "A red and black bear came to help me hold my arms up," something that had been a challenge for her in this posture.

For others that challenge was to let energy flow through them.

Continuing with Jessica's story: "Energy from the sky was coming into my mouth and chest area, but could not flow because it got stuck in my hands. . . . I wanted to open them up to release the building energy, allowing it to flow out. . . . The feeling of being stuck made me angry and I didn't like the battle between my top and lower body. I felt feathers on my arms but couldn't fly because I felt I was cemented to the ground." Another group member reported being encased in concrete. "It was as though I was encased and trying to break out. I wanted to fly but couldn't. Finally I felt encouraged to let go of my frustration and I flew over mountaintops. The intense heat in my head and on my face felt the cool air." The top of Joan's body swayed while her lower body was rooted, causing her to feel like a tree in the wind. Donna was told she had many bad ideas stuck in her head, and she experienced spinning and whirling that served to clear her upper body. For Bruce, the lower half of his body "collapsed" into the upper half and then he felt light, open, and weightless. The sun was a common image. One woman saw a figure with a red sun as its head. Another was drawn to the sun and was praying to the Sun God until she eventually became that Sun God.

The overall effect of this powerful spirit journey is summarized by one man's report: "The power of the heart is rooted in the sacrum. Open the sacrum and the throat, allowing connection and alignment for the flow of sun energy or divine energy." Now it all made sense. The journey this posture facilitates is the one between Earth and Heaven, which occurs in the human body: it is the rising of the kundalini, the snake body that moves up the spine through the two intertwining channels, through the opened sacrum, the heart and the throat, to fully activate the crown chakra or the internal sun. The kundalini, or *sekhem* in the Egyptian system, is portrayed as a winged solar disc.[1] The winged serpent is the caduceus, the familiar symbol for healing.

I returned to an earlier experience of my own with the Bird Woman of Egypt. In an ecstatic trance, I had seen a huge snake coiled around a nest of four eggs. In the Hindu yoga teachings, the *muladara* chakra at the base of the spine has four petals. Its colors are red and black. Within this chakra sleeps the kundalini *shakti,* the great spiritual potential that lies in the form of a snake waiting to be aroused and brought back up to the source from which it originated, Brahman. "Don't abandon them," she told me. They were cracking open and dark wet birds, feathered serpents, were emerging. "Hold on to the body of the snake," she said, "and become Her: long, sinuous, female, and powerful."

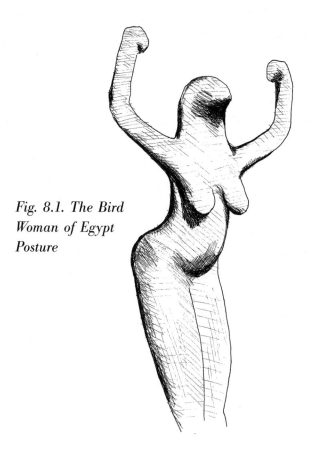

Fig. 8.1. The Bird Woman of Egypt Posture

Performing the Posture: Stand with your feet close together and your knees locked. Arch your back and fully extend your neck. Make fists with both hands and raise your fists above your head with your arms spread apart as far as you can hold them. The palm of your hands should be facing each other, rather than forward. Raise your chin so you can look at the sky or ceiling above you, and close your eyes.

The Sleeping Lady of Malta Posture

Hal-Saflieni, Paola, Malta, is an amazing subterranean structure that was built around 2500 BCE. This ancient necropolisis the only prehistoric underground temple in the world. Discovered in 1902, the Hypogeum Temple was accepted by UNESCO as a World Heritage Site. Its name means literally "under the earth." On the first level, the oldest level, rooms were made by extending natural caves that were then painted red with ochre, a common practice for early burial sites. The small clay statue of the Sleeping Lady was recovered from the main chamber surrounded by offerings and is now held in the Museum of Archaeology, in the capital city of Valletta, Malta.

There is speculation that she represents an important woman who had died and was buried there, but our experiences indicate that she can lead us on a spirit journey to the Lower World, perhaps the place where her spirit resides. Her ample figure more likely reflects the hills and valleys of the Earth than the characteristics of a particular person. A few of us may have met the Sleeping Lady in her waking or living form and in these cases she was not unusually large. "I saw a dark-haired woman with a white veil, like a bride, carrying a small bouquet of dark purple and crimson flowers. She seemed ordinary. Then I was the woman and the veil was created by the shower of light or water under which I stood."

Susan described "tunneling down through purple and dark blue hazy swirls. . . . I was then looking through a tube at the parched earth or skin and light glowed through the cracks as if it were map lines of countries on the earth." In other words, her journey took her to a new place. "I was then lying down and hundreds of foreign beings encircled me looking at me, wondering who I was, curious as to why I was there and how I got there." Her journey continued, introducing her to strange animals and shadowy figures, and then everything faded into a pale green light.

Bruce told about being drawn into a circular counterclockwise vortex before passing through a thin layer above him and hovering in what he identified as the dreamtime or the astral plane. He felt suspended in a parallel place. He asked why there was no movement in this spirit journey and was told that his intent, for the trance that evening, required no action or travel.

While Michele reported being relaxed and rejuvenated, John found the experience disturbing at first as he descended to different levels, finding first an owl looking at him, then a tiger pacing, then people from the past who were long dead, and finally other animals. Toward the end of the fifteen-minute rattling session, he started the ascent to the world of his normal life. Another group member was also unsettled by the posture and said that his intent was to protect himself, to "set boundaries and not let anything in." He reported that the posture was very comfortable but nothing much seemed to be happening. He was challenged to step into the experience but continued to be reluctant to do so. His only experience was "an instantaneous sensation of falling or dropping, but my consciousness was closed off to it."

In a journey to the Lower World, it is not uncommon to encounter those who have died, and this association with death makes the journey uncomfortable for some. In my trance, the floor beneath my feet kept shaking as though someone were moving. I opened my eyes and saw the group peacefully reclining, no rocking or shifting. Then I saw Her, the figure, impatiently thumping her leg to get my attention. When I finally focused on her she told me to come to her. I asked, "How shall I approach you?" She said, "Come to me naked, on your knees, in the darkness." For me it was an invitation to face the darkness of the Lower World with her. A few months later I enacted her ritual in the moonlight on the small porch of a cabin in the forest. It is only because of many, many trips to the Lower World—entering through diving into water or falling down waterfalls or into deep holes in the earth—that I am finally able to approach it with only minor flutters of anxiety. The lessons I continually learn are about the inevitability of death and an invitation to learn her secrets so that when it is time for me to literally die I will know my way to the place where the spirits of those who have died may dwell in peace. This may be the gift of the Sleeping Lady of Malta.

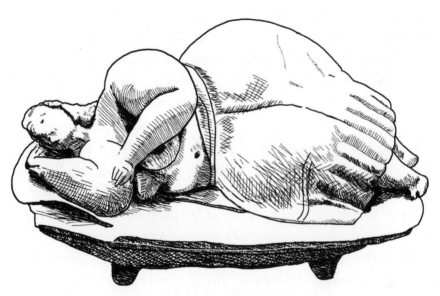

Fig.8.2. The Sleeping Lady of Malta Posture

Performing the Posture: Lie down on your right side. Bend your right knee and place your right foot in front of your left foot. Bend your right arm at the elbow and bring your lower arm in front of your body so that your head can rest on your right hand. Fold your right hand into a loose fist. Let your left arm lie on your upper body and bend your arm at the elbow so that your left hand can rest on your right elbow. Keep your eyes closed while facing forward.

The Poppy Pod Woman Posture

Described by Marija Gimbutas as a young Earth Mother, this figure was found on Crete and is from the late Minoan period, dating from approximately 1350 BCE. The headdress she wears is decorated with poppy pods that archaeologists identify as the source of an opiumlike substance that induces sleep, although we believe the "sleep" associated with her is a trance state. Gimbutas writes, "She is the same goddess who appears on seals and signet rings seated beneath her tree, receiving offerings of poppy heads and flowers, symbols of her bounty."[2] There is ample evidence that those who call upon the Poppy Pod Woman through this ritual posture experience the presence of a rich and bountiful earth who provides energy and sustenance to her children.

Our first experience with the Poppy Pod Woman was on the occasion of Felicitas's eighty-fourth birthday, in 1998. A group of eight women participated in the trance. Without any idea what the posture might mediate, everyone experienced movement and energy, with a few people directing the energy toward healing others. Literally, every person in three different groups who experimented with the Poppy Pod Woman Posture reported a distinctive energetic movement during the trance. Michele described it like this: "Very oddly through the majority of the posture I felt something pushing me back through my abdominal area and I just kept swaying back and forth. I tried to stop it but then I would get a greater sensation pushing again." Joan called it a very physical trance, with her arms and body dancing. Someone else saw the Poppy Pod Woman doing a slow hula that began moving faster and faster, twirling and rotating like a corkscrew. The twirling caused a golden gauze to form at the center of the circle and in the midst of the swirling she said she had "one of the most ecstatic energy experiences I have ever felt."

There were several reports of an entrance in the center of the circle. Chris saw a black and purple hole that she entered. The hole became a long hallway that opened onto a scene at the other end. She said, "It was a doorway to another world." Carol also saw darkness in the center of the circle "that I could have dived into" but she chose not to. In Linda's trance she felt as though she had fallen down a hole and said to herself, "I'm falling down the rabbit hole," referring to Alice's introduction to Wonderland. Jackie felt something sucking at her feet and a sense of being pulled into the "belly button of the Earth." She was tumbling down in a spiraling motion until she was suspended in the midst of bright cores of pulsating light. "I could touch them with my hands and the current fed my soul."

John fell into a pool of water and felt refreshed as the waves of water continued to flow around his body. In front of him were green and blue lines. When he asked what they were, he was told they were the earth energy lines that are all around us, ley lines that create a matrix around the planet. It was as though Earth was showing him the network of energy that vitalizes the planet while giving him a personal experience of that vitality as refreshing waves in a pool of water.

Other people encountered the Earth Mother in the form of trees and stone: being in an old-growth forest, or flying over first a desert landscape and then green forests with huge trees and a river flowing through it. There were branches that might have been antlers but became tree bark, fields of poppies, and sunflowers. Jackie said, "I became part of the ecosystem and I saw all of us, it seemed like all females everywhere, with streams of energy flowing out of our hands like a meteor shower." Bruce saw an African woman with energy flowing from her hands and then saw an image of growing plants, leading him to believe that she was teaching him about how to make plants grow.

In Claire's trance, she saw a tall tree with mirrors in its branches. A band of monkeys threw the mirrors on the ground and broke them, but the fragments were magnetic and they pulled themselves together again. A stream of deep red water, the blood of the earth, flowed into the tree and into the mirror pieces. Each mirror reflected a hand with an eye in the middle, an old symbol for the eye of God. Her story might be interpreted as an archetypal tale. Throughout nature, the Earth Mother gives us humans the opportunity to see ourselves, to know ourselves in the reflection of trees and stones and sky. With our monkey minds, we tear at these delicate mirrors, wanting more to control nature than be taught by her. We throw the gifts to the ground and break them. It is our salvation as humans that the pieces can be put together again, even without our help, and the deep red blood of the Earth Mother restores all of nature, including ourselves, to wholeness once again. When we are able to see, each branch and leaf shows us the all-present, all-seeing eye of the creator. Looking in those mirrors we can see that we are not separate but one of innumerable cells in the body of the Mother, tiny drops in the ocean of her being.

When Judy asked the Poppy Pod Woman if it would be possible to present questions to her, there was no answer, and nothing about this group's trances suggested divining. In my own trance I was directing healing energy toward my father, who had been struggling to recover from several surgeries and an ensuing infection. I saw him being swung in a hammock as a baby would be soothed and comforted in a cradle. Judy asked for healing for another woman and "felt myself get taller. A white light was coming from the top of my head. I thought it was a little owl [a spirit friend of hers] and I was very happy." Nevertheless, the combined experiences of several groups suggest a very special kind of spirit journey through the subtle energy matrix that supports the earth and each of us, a journey through what might be called the Middle World. Through these experiences, we find ourselves restored and can even share this bounty with others.

Fig. 8.3. The Poppy Pod Woman Posture

Performing the Posture: Stand with your feet parallel and about six inches apart. Keep your knees slightly bent. With your arms at the side of your torso, raise your hands with your palms forward to a position above your shoulders but below your ears. Hold the fingers and thumb of the left hand close together. Hold the fingers of your right hand together also, but keep your thumb separate from the fingers on this hand. With your eyes and mouth closed, face forward.

The Tsimshian Shamaness Posture

Identified simply as "a woman shaman," this wooden carving from the Tsimshian tribe on the Northwest Pacific Coast is housed in the Provincial Museum in Victoria, British Columbia. The shamaness wears a curved bone through the septum of her nose, and a large raven sits on top of her headdress. Both of these characteristics identify her special status as a medicine woman.

From our earliest experiences with the Tsimshian Shamaness, we understood that in trance she takes us on a spirit journey. After our first journey, Felicitas wrote, "This is the world where the shamaness is descending to look for friends or helpers." Curiously, Felicitas's report was the only one in nearly eight years of experience with this posture in which animals appeared. Animals indicate a journey to the Lower World where, according to the traditions of indigenous peoples around the world, the animal spirits live. However, in this posture, most people were greeted by birds of every variety, Crow or Raven as suggested by the shamaness's headdress, Parakeet, Cockatiel, and Parrot, all indicating a spirit journey to the Upper World. Elizabeth flew with Raven, beautifully describing the experience as "rowing a canoe with Raven in the air," and by looking into "his very large black iris, I could see the map of the universe." Jan saw her friend's Raven fetish flying high to create a sky bridge, "pathways made of ribbons of light" that made it possible for her to travel to the Realm of the Dead. Olga reported Dove, Hawk, and Falcon, and Joyce saw "eagle feathers way up high."

In Joan's trance, she saw Crow come out of the rattle and she herself grew big black wings. Even though she flapped her wings, she did not go far. Similarly, I was invited to fly with a parakeet but became a bird covered in oil—and so unable to fly—and in another trance saw a rooster, a bird that does not fly at all. These experiences all point to the difficulty that we sometimes have in leaving our bodies to fly in the spirit journey to the Upper World. Being tired or preoccupied with other concerns can interfere. However, on a different occasion, my head had a funnel on the top, like the tin man in *The Wizard of Oz,* and I shot out into the sky. Bruce described getting out of his body as a "swirling spiral column moving up and heavy gravity pulled me." In Chris's trance experience she saw a male figure change into a pure purple energy that shot up through a translucent tube that pulled the energy upward, creating a path for others to follow. David was in a cave and "the opening of the cave was shaped like the top bone of the optical orbit of the skull," highlighting the top of the head as the appropriate exit for the spirit in journeying to the Sky World. Others described going up a tall staircase or being at the top of an obelisk that "opened from the top like a portal."

The heights of the Upper World are sometimes characterized as mountain vistas or mountain peaks covered with snow. However, the change in perspective may sometimes be less dramatic. Jackie felt like she was a reed in the water, swaying gently and watching the people and animals come to the water to drink. When a moose came by, to eat the reed, Jackie was pulled up and could see from a higher perspective, which is one of the primary reasons to undertake a spirit journey to the Sky World.

Whereas most people seem to travel to the Upper Worlds with the Tsimshian Shamaness Posture, MaryAnn was told that in this trance "you can go anywhere you want." Felicitas journeyed into her personal past for visions from her childhood to enhance her writing of her autobiography, and Dominic traveled to what seemed to be an earlier time, to a gathering of Native American tribes. There had been a betrayal of an agreement, and as a result, the earth was scorched by fire and bloodshed. By traveling in time, he returned to the earlier gathering to view it from a higher plane so that the thoughts of every person and their intentions could be clearly seen and the betrayal thereby revealed.

Most people report significant swaying, rocking, pulsating, vibrating, or twitching, indicating the level of energy being built up to accomplish the spirit journey. Jackie said that "bolts of energy jolted me a few times," while Rae was a little more removed, seeing lightning in the background. Usually we are gently escorted into this amazing capacity to travel into other realms of reality, but sometimes the full impact of becoming "a journeyer or a wanderer" hits us and we experience the loss of familiar ego. Once the shamaness told me, "You do not know our ways and you will not like them." Later she appeared to me during another trance and I asked her what this spirit journey was for, how I could serve, what I should "take back to my people." Without a word, she inhaled me. While I was in her nose, she said, "You can serve best by changing what you are," conveying clearly that my best service is the transformative work I do on myself. Rather than feeling humiliated— after all, she had just snorted me up and was holding me in her nostril like a piece of debris—I had to laugh at how pompous I was being in asking those questions. Like a no-nonsense grandmother, she put me in my place without pushing me away, and because of that, I will go back to her to learn about relinquishing my ego attachments. When we can live in constant connection with the spirits, they can live in the world through us.

Performing the Posture: Stand with your feet parallel about six inches apart and with your toes pointing straight ahead. Bend your knees slightly. Hold your right arm close to your body and bend it at the elbow so that your lower arm extends forward at waist level. Hold your thumb and fingers on this hand close together and allow your hand to droop from this position. Hold your left arm close to your body and bend it at the elbow to draw your left hand all the way up to your left shoulder. Keep your thumb and fingers on this hand close together and allow your left hand to rest on your left shoulder. Allow your mouth to droop open. Keep your eyes closed as you face forward.

Fig. 8.4. The Tsimshian Shamaness Posture

9

Initiation Postures

Spiritual initiation is a process of learning, step-by-step, how to die and be reborn. How wise these ancient cultures and civilizations were to have a well-developed method for teaching everyone this essential skill that puts all the rest of living into a meaningful context. For many years I have thought about ways to incorporate initiation postures, and the skill of learning how to die and be reborn, into contemporary workshops. Helping people to know what to expect can only assist in learning to easily let go at the time of death, and understanding what is occurring is a wonderful support for the friends and families of those who are dying.

A decade ago we organized a workshop in Missoula, Montana, on death and dying. Missoula is the home of the Chalice of Repose project in which musicians are taught how to accompany people in their dying process, using music to ease the transition into death. Graduates of the Chalice of Repose project as well as the local hospice organization attended the workshop and were very enthusiastic about the valuable addition of initiation trances to assist those who are dying. A board member of the hospice organization said that every staff member and volunteer of every hospice program should have this training, but at the time we did not have the resources to follow through with this vision. A few years later two of us offered workshops for cancer patients and their families using initiation postures; the group members loved it but the hospital staff was uneasy about our unorthodox method. We were not invited to continue.

Most recently I returned to Santiago, Chile, last winter to continue teaching ecstatic postures. We designed a workshop called Buen Morir, or Good Dying, to teach people how to learn about dying through trance and then to become "midwives" to the dying. While hospice is a program of palliative care for people at the end of life, Buen Morir gives their friends and families the tools and resources they need to support this end-of-life process in a loving and meaningful way through understanding the losses experienced by the person who is dying, how to talk about dying, completing end-of-life tasks such as saying good-bye, clarifying one's legacy and making meaning of one's life, making plans for rituals, and managing pain through healing postures. We teach breathing exercises and go through an experiential process to replicate the loss of roles and identity that causes so much anxiety when people approach death. Through the trance experiences, workshop participants began to discover what Kathleen Dowling Singh identified, that "dying, remarkably, is a process of natural enlightenment."[1] My colleague, Paula Olivares, has facilitated another Buen Morir training for a hospital staff in Santiago and we hope the program will continue.

The first time I experienced dying through an ecstatic trance posture, I found myself traveling down into the Realm of the Dead, falling like a rag doll down a steep bank. Later I learned that this was like a spirit journey to the Lower World and I came to anticipate going down, sometimes as a skier or maybe diving off a high cliff into the sea. When I came to a fire, I entered it willingly and felt the muscles on my bones grow slack and then drop away until I was only a skeleton. Oddly enough, I observed all of this dispassionately because the essence of "me" was not in the muscles or bones, but was simply a curious witness. Supported by the posture and the sound of the rattle, I waited quietly as my bones were placed in the fire, the final release of the molecules and atoms I had borrowed to make a physical body. It was quite dark and there was nothing to do but wait. My spirit began rising above the earth, into pink and blue clouds, then upward toward the red disk of the sun. The sadness of saying good-bye was replaced by tears of joy. Elated, I knew I was going home.

This group of initiation postures is in female forms and all of them come from the Middle East and Egypt. The first one, the Inanna Posture, supports a journey of descent similar to Inanna's journey as told in Sumerian myth. The experience facilitated by the Ishtar Posture focuses on sensuality, sex, and fertility. This is the stage in birth, death, and rebirth in which we take on a physical body and learn to enjoy it. The Sekhmet Posture teaches us to be awake in death and rebirth, and to learn the dance of coming into form, landing in newly acquired bodies, and then leaving form. Finally, the Shawabty Posture seems to focus on how to die and often takes us into experiences of funerals and similar ceremonies, so that we become familiar with this part of the process and do not hold back from it. Paradoxically, learning how to die helps us to appreciate what it means to be alive and how to fully live.

The Inanna Posture

The Sumerian myth of Inanna, Queen of Heaven and Earth, is written on clay tablets from the third millennium BCE, at least five thousand years ago. The myth tells the story of Inanna's descent into the Underworld to rescue her lover and consort. There, in the presence of her dark sister Erishkigal, queen of the Underworld, she is "stripped bare . . . judged by the seven judges . . . [and] her corpse is hung on a peg, where it turns into a side of green, rotting meat."[2] Inanna survives her ordeal with the aid of Enki, god of waters and wisdom, and returns, giving us a classic story of initiation into death and rebirth.

In a series of research sessions over five years, we experienced these cycles of death and rebirth, of descent into darkness and the eventual return of light, of decay and rotting that finally became flesh, blood, and bone. It begins with the "peeling away of all but the essential," as Ruth described her experience, a parallel of Inanna's being stripped bare, followed by the darkness. In John's trance, he reports, "Then things became black. I saw nothing, heard nothing, felt nothing but blackness. I was beginning to think that was all when light began to flicker." Others spoke of dark shadows taking on a rainbow aura or a shift from gray to color. The cycles of death and rebirth were most dramatically told in one trance in which the hawk killed the dove, the dove fell to the earth, but a tree grew out of the earth and the tree was filled with doves.

This is a very feminine posture, voluptuous and fertile, leaving Michele with the feeling that she was proud to be a woman. In one session we were a group of only women and I saw us all with large bellies: "we were all pregnant with the earth, all blue and green in our wombs. We were standing in for the universe, birthing the earth . . . linking the mystery and beauty of fecundity among women across time." One woman saw many images of fertility: belly dancers, Kokopelli seeding the ground, a group of young women dancing around a drum with flowers in their hands. Others reported lush flowers, like the "huge red flower with its pistils twirling." One woman said, "I was Mother Earth," and Rebecca saw herself become a goddess sending streams of light over land and sky. Another woman described seeing a woman like a mountain with blue light coming from her in waves, her orifices exuding water or milk.

This image of milk flowing from the breasts of Inanna as well as the women in trance initially led us to interpret this as a healing posture. We used it as such on two occasions, to send healing to Kosovo in April 1999 and to Chechnya in April 2000. The results were a global sense of empowerment, an outflow of strength, courage, faith, and love, a deeply feminine sense of empowerment. Here is a sample from one woman's trance: "I saw a landscape at the edge of a town, with dark hills or mountains in the distance. There was a little sunshine. A woman and her child were huddled against the remains of a building. She had gathered straw to help them be warm and the wall broke the force of the wind. I felt the cold, the physical sense of not having bathed for a long time, of being hungry and afraid and on the edge of despair. We [in the trance group] were like spokes on a wheel, connecting to a central fire, the energy from our own breasts feeding the fire. In that other place, it grew dark and people gathered around the fire and danced, raising their spirits, blocking out the hunger and fear. Then the center became a fountain that anyone could drink from. The spirits danced around us here, weaving the energy we each directed, helping us too."

Several years and many sessions later, it seems clear that rather than specific healing, trance with this posture provides a deep sense of renewal, empowerment as primal as the budding of new leaves after a cold winter, the strength of rebirth even in the face of death. Initiation includes this capacity for coming back and is why we always include birthing postures in the category of initiation. Even men experience giving birth, although often less directly than women. David reported seeing an etheric feminine form with "worlds being born through her vagina," and John met a large black woman who was "testing me to see if I was afraid," but later touched him reassuringly to let him know that he would be fine.

In experimenting with the stance, I found that the trance was stronger for me if my fingers touched. Niki was told to push out her stomach and lower abdomen, and then "there was a whoosh of energy up my vagina and out my breasts." Many people felt energy coming up either through their feet or their vaginas, causing them to have physical sensations ("my back cracked") or the experience of "opening up" or a desire to leave the body with spasms in the thorax and pelvis. Victoria was guided to perfectly balance the weight of her body on her two feet and then a column of light came up the center of her body. For another the rattle split and there was a rod of energy where the rattle had been. The splitting was replayed again and again, into light and dark, into a balance of remote and immediate, of ordinary reality and the cosmos.

There were very few animals reported in these trances. In fact, many people commented on the absence of animals and the presence instead of humans. Felicitas saw people streaming toward her and Chris saw "lots of people milling around, talking, singing, trying to tell me something." While Niki saw dark-skinned shadow people, others saw them with auras of white or gold. Many birds were reported, especially Owl, and in two dramatic stories, on different occasions and by different women, a snake engulfed each woman. Linda described it like this: "I see snakes slithering around, their tongues flicking out. . . . I feel my own tongue do the same. I feel it flick out and move like theirs. Then I no longer see them but feel a large snake wrapping itself around my body, beginning with the feet and moving slowly around. I am not afraid of this: I only observe." Similarly, Mary saw herself standing next to a cobra that began wrapping itself around her in a very sensual way, holding her while the earth trembled and there was darkness and pain all around. Much later I found a continuation of the myth of Inanna, in which her consort is transformed into a snake to escape from being seized by demons from the Underworld.

It is not uncommon for people to ask for personal help in any trance and so we find reports of advice and admonitions that continue the theme of empowerment in the face of even death: "Wherever you go, there you are," "Stand tall and straight in your power, in who you are," "Move through your fears and you will always be you."

Fig. 9.1. The Inanna Posture

Performing the Posture: Stand with your knees locked and your legs held close together so that your knees are touching. Cup your hands under your breasts with your fingers held close together and the fingertips of each hand touching, and with your thumbs circling the outside of each breast. Relax your shoulders and upper arms, so that your elbows extend slightly away from your body. Close your eyes and mouth and face forward.

The Ishtar Posture

This delicate figure of the Babylonian goddess Ishtar is over three thousand years old and now stands in the Louvre Museum in Paris. She is carved from white stone—probably alabaster, the medium for a number of important statues from that era—with a precious red stone, perhaps a garnet, adorning her navel. Both arms appear to be attached to the body on hinges to allow for the fine articulation of the hands and arms. Her elaborate earrings and coils of necklaces around her throat are made of metal and on the top of her head is a small half moon. Ishtar has been associated with Venus and that little half moon makes her resemble the astrological icon for the planet Venus, a symbol recognized around the world as representing woman or the feminine.

Ishtar might be considered the younger sister of the Sumerian goddess Inanna, and, like the posture of Inanna, this one is also useful in experiencing initiation, although not so much about death and dying but more from the perspective of fertility and hence rebirth. In the Inanna Posture, we seem to have experiences that focus on personal power, whereas the Ishtar Posture facilitates opening to grace, beauty, and wisdom.

Ishtar's priestesses were probably sacred prostitutes, who in the sexual act embodied the goddess and celebrated the mystery and beauty of the perpetual renewal of the earth. In trance, women especially experienced luscious sights, smells, and physical sensations. Mary saw a majestic temple with steps and columns, stained glass, crystals, and a floor inlaid with gold and entered an open space of white floors, blue pools of water, sweet smells, and delicate music. For others the beauty extended into the natural environment, as they described lush fields of wheat, blue skies, undulating brown mountains, a stream bordered by weeping willows with butterflies and hummingbirds creating flitting dots of color. One woman described her own voluptuous body being massaged with fragrant oils, and then saw the group of women standing in a circle in trance, "looking so lovely that a sweetness came over me." Bruce saw "women and girls behind veils and curtains," and Felicitas described being immediately naked, like the statue, "naked and ready," suggesting women prepared for their role as priestesses of holy sexuality. One man experienced being with a priestess-become-goddess: he saw a woman's legs, her ankle bracelets in gold, the scalloped fabric of her clothing. He was mesmerized and followed her to a bed decorated with plumed plants and scented with oils. After their love making, he saw her select another man, who followed her in exactly the same way, "magnetically pulled to her."

Judy's message for her friend was one of "sexuality and sensuality" and she commented on her own feminine receptivity: "this trance makes me more accessible for receiving than other trances." Another Judy said the center of her body was open and radiance poured through, and in another trance "my molecules separated and I disappeared and became one with the light." Indeed, for many there were streams of light or fire, dissolving into swirling energy. I commented once that there was no masculine energy here—no goals, no tasks, no disciplines—just "sweet roses blooming in my heart; they are all that I ever need, expressions of the Mother" ever present in our lives.

Sex and fertility are part of the cycle of initiation, the beginning of new growth and fruition and inevitable dying, each phase a melody in the sacred song of life. The focus in this posture is clearly fertility. Many people have described participating in fertility rituals. "The left hand was used for bringing energy up from the earth and the right hand used to direct it," "I pulled energy from the ground through my left hand and it glowed as a ball in my right hand," "I felt like a big seed and the rattle exploded seeds into the earth, falling on the earth and she responded," "there was blood on the ground and we were all bleeding," fertilizing the ground. Snakes, a representative of the great cycles of transformation, appeared everywhere and sometimes in a stylized form of spirals or as undulations up the spine. Claire saw a "large black snake, weaving hypnotically, that then became a tree with lots of little snakes in it, falling to the ground." A pregnant energy materialized in the right hand of many individuals experiencing the trance state. Some described it as a ball of energy, mostly warm, heavy and soft, or a ball of energy from the sky or a globe, "the ball of the earth in blue and green held in the palms of our hands." One woman saw a great tree—like the Tree of the World—materialize in the center of the circle and absorb all the balls of energy from each person. Although there was always a lot of energy, it was well grounded in the earth, even through a pole of energy in the hand of each person that kept her attached to the ground while still leaving her free to move about.

The experience of being one of an intimately connected group was repeated from one session to another. "We were all angels with wings, flying around Earth, gently touching our wings to help us keep our balance," "there were women together singing and grinding corn," "there were women dancing in circles." Jackie described a spider mixing our hearts and brains into packets. "We eat the packets and there is a sharp connection of knowing and feeling each other from within."

Some described the presence of warriors, marching around the base of a rosy obelisk to the beat of heavy drums. One woman said, "We were trying to hold our own amidst the forces of war." Among the Greeks, sacred prostitutes were especially important to warriors returning from battle who would visit the temples to ask the goddess to balance the aggressive violence of war, preparing them to return to wives and children and the cooperative values of village and family life. This quality of healing appeared in a few trances, as Claire described healing herself from chronic pain and me from my weariness.

Animals appeared briefly as guides and helpers. In addition to snakes, there were many jaguars, several bears and eagles, a "she-lion" or lioness, and a doe.

Because this posture facilitates trance experiences that are so feminine in nature, Dominic raised the question: Is this posture just for females? He was told that his intention could be to investigate, that the posture was used by initiates for the purpose of receiving training, and that he could participate. Moving through the lower astral levels—"through ghouls and ghosts"—he arrived at a bright city in the clouds. There he found himself in a room with teachers and received instruction. Upon his descent and return, a group of women who had been his classmates draped him in a robe and he withdrew to digest the information he had brought back with him. At that point, he realized he had been a woman. The teaching reminded him of Sophia and the love of wisdom, or *philosophia*.

Some people are more inclined than others to hear verbal advice or instructions. The wisdom in one message was simple and profound: "Be in your body, love your life, don't hold back from your experience." And I can still see the radiant tears in the eyes of the woman who simply said, "She called me Sister."

Performing the Posture: Stand with your feet parallel and about two inches apart. Keep your knees slightly bent. Hold your left arm to the side of your body with your elbow locked. Keep the fingers of your left hand close together but extend your thumbs away from your hand and point your fingers toward the ground. Bend your right arm at the elbow and with the palm of your hand up, hold your lower right arm in front of your body at a ninety-degree angle. Keep the fingers of your right hand extended and spread apart. Close your eyes and mouth, and face forward.

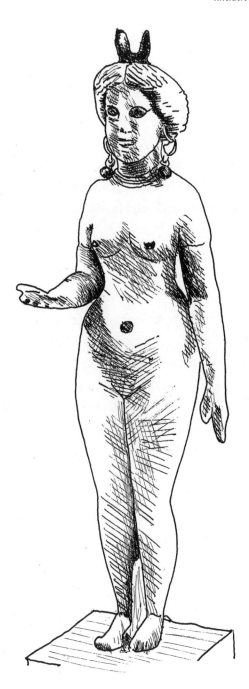

Fig. 9.2. The Ishtar Posture

The Sekhmet Posture

Be sure to stay clear about why you are going to war: what is the purpose?

Be sure to remember that you are not going to battle because you feel like fighting.

Strategy, focus and purpose are everything.

If you must fight, stay in your body and fight the good fight.

These words were the advice of Sekhmet, the Egyptian goddess with the head of a lioness, when we first consulted her in trance several years ago.

In his book *The Goddess Sekhmet: The Way of the Five Bodies,* Robert Masters writes, "The Goddess Sekhmet is undoubtedly one of the most ancient deities known to the human race—much older than Ptah, her brother-husband, or Ra, her father, the sun god and King of the Gods. . . . The famed classical Egyptologist Sir Wallis Budge comments that the name Sekhmet has been derived from or is connected with the root *sekhem*—to be strong, mighty, violent. Others have added the meaning of sexual power and it is believed that Sekhmet is the Goddess of the Kundalini energy constellation."[3] While many writers emphasize Sekhmet's violent and destructive nature, we discovered that her more powerful role seemed to be as one who initiates humans into the mysteries of death and rebirth.

Masters identifies five layers of "bodies" as an ancient Egyptian teaching associated with the priests of Sekhmet. The first body is the physical one that functions nearly automatically to maintain itself as an independent, self-renewing vehicle for consciousness. Ba, the second body, is composed of the mental and emotional images we hold of ourselves and is what we usually refer to as self. The third, the Haidat, holds the personal unconscious and in some systems is referred to as the dream-body. The fourth is known as the magical body, because when we are conscious in this body, we are capable of manipulating physical reality. At this stage of spiritual development people can acquire powers or *siddhis* (as they are known in Sanskrit), and the religious literature from every tradition is filled with stories of miracles performed by saints and medicine women, the holy ones and the religious specialists. Finally, the fifth body is the spiritual body, the body of enlightenment where we reside when we witness the truth of unified reality and can participate in the world as a translucent shimmer of consciousness.

What is death and rebirth, after all, but the shedding of old bodies and appropriating new ones? The curious term *dematerialization* came up in several different trance sessions using this posture until it became clear that through our experiences with Sekhmet the cycles of taking on form and dematerializing form were being explored. "A humming circle of light moved around my body. It hovered six or eight inches around me, going faster and faster until I was held within a cocoon of vibrating circles of light that dematerialized me. Then slowly from the top of my head a single circle manifested a physical body for me."

The role of consciousness, of being awake in this process of death and rebirth, was emphasized in several ways. Bruce was in a chair at the center of a bubble craft and was learning how to use the hand controls to guide the craft. Alice, after being escorted on an extended journey through the materialization of the planet, was handed the keys to the sacred doorway and her guide disappeared; she knew the mystery now, and was on her own to use it as she chose. Darryl saw a figure dancing around a fire, then he jumped into the fire and through a long tunnel of fire until he popped out into a quiet void. He, like Brian, entered a vehicle and through the screen witnessed the vast starry nothingness. Later the void became light, without movement, without space or time, with "infinite access to infinite information, without the need to move because I was connected to everything."

The cycles of materializing were displayed in a range of venues.

Ed saw the rattle at various stages of its development, beginning with the subtle plane from which the inner pebbles and seeds came. Anne went through the initiation journey as the body of the planet took on form: jewels and minerals, coal and oil, wheat, sand, ocean, and waterfall. Chris also went to the core of the earth and found a new body for herself there. "Something scanned my body, stretching me, making me—my aura, what holds my energy—bigger." It worked on blockages in her body to make her aura and her brain bigger to be able to hold "what I have to do." In this place there were "a ton of people milling around, waiting, not doing much," presumably waiting for similar adjustments for themselves.

Brian acquired more integration in the layers of his mental-emotional body as, "out of the fog," he saw himself at eleven years old as well as the adult version of himself and the "parent self" informed by his biological parents. Much of the trance session was spent helping them learn to get along, while at another level he experienced himself as a warrior in another life. He had a sense of learning to be a warrior when there was no war, how to be a warrior and to be peaceful at the same time.

Consistently people reported feeling centered, awake, and empowered, making conscious choices and viewing their bodies as newly acquired resources. Several guides showed up from specific previous trances, human or composite forms instead of animal allies. Although a number of people experienced a journey, it was more a story of the cycles of leaving form and coming into form than the traditional spirit journey. The advice that Sekhmet initially gave to us still holds true: "What is the purpose? Strategy, focus, and purpose are everything. If you must fight, stay in your body and fight the good fight."

Fig. 9.3. The Sekhmet Posture

Performing the Posture: Sit very erect in a chair with your feet parallel on the floor about six inches apart. Make a fist with each hand and rest your thumbs on top of the fists. Place your fists on your thighs, just above your knees, with the thumbs on top. Keep your elbows locked. Close your eyes and your mouth, and face straight ahead.

 # The Shawabty Posture

Toward the end of the late Middle Kingdom in Egypt, about 1900 BCE, a practice developed of creating wooden figures from the persea tree, known in Egyptian as *shawab,* to be placed in the tomb of one who had died. The purpose of the *shawabty* figure was to accompany the deceased on the journey of the afterlife. Sometimes called "the answerer" or "the one who responds," the shawabty was to be a substitute when the gods called upon the deceased to perform hard labor in the fields of the Underworld. Each shawabty was equipped with a hoe and basket or other implements of labor as well as a spell that ensured hard work and obedience. Whereas most individuals had one or two shawabties, 413 were discovered in the tomb of Tutankhamun.

Although the earliest shawabty figures were made of wood, later ones were formed from stone, metal, terra-cotta, or faience, a non-clay-based ceramic that was used in Egypt as long as five thousand years ago. Characteristically they have youthful features and a peaceful smile, but they are primarily identified in a standing posture with arms crossed over the chest. During research sessions with this posture, several people were instructed during the trance experience to keep their feet together with their ankles touching and to keep their knees stiff. In this pose, their bodies seemed to sink into the earth as though the feet and legs were stakes pushed into the ground.

From the beginning of our investigation of this ritual posture, it was clear that we were being instructed in the ways of dying. Susan reported, "As soon as the rattling started, I felt a cool breeze and cool blocks of porous stone going down into the darkness. I was dying, just dead, and descending into the tomb. I was afraid. I thought maybe the shawabty figures were buried with the dead to be companions. My husband appeared on my left and my Corn Mother fetish on the right. Then the Corn Mother assumed the shawabty pose and so did my husband. Nothing to fear." Susan then went toward the light she saw shining just beyond her and there she saw her mother who was dead. Her mother's face was smiling and peaceful just like the faces of the shawabty figures, and she reassured her that dying was not so difficult.

Janis saw a dry, sunny, barren land, then flooding, "as the Nile would flood." She summarized her experience as being about destruction and renewal. She heard the words, "Come home," and she knew she was being taught about death. Rachel was in a mummy casing and saw animal spirits darting around her. She was being prepared for the death experience with a necklace and her head was placed on two straw disks. She walked into a cylinder that enclosed her and there her body began to spin. She felt hot, then cold, then hot again, repeatedly experiencing that fluctuation. Eventually she dropped her physical body and the cylinder opened, revealing only a pile of sand. Darryl also experienced twisting, "like digging, for a burial." He discovered that when he held his hands close to his chest, the twisting diminished.

A number of people experienced some form of a funeral procession. Michele heard within the rattle the sound of horses trotting and experienced being a large statue moved by horses, swaying along a bumpy road. Linda was being pulled by a turtle on a flat conveyance underwater. John was on a boat floating through a lush, green environment that he took to be the Amazon River. He said that they were traveling to a ceremonial place, a dark circular hut where there was chanting, "many voices all around me," in a ritual related to the stars. Another man, also named John, saw the water below and the jungle above, filled with a chorus of birds. "I was the statue, a female, and was being placed in the water to become a guardian of the temple, where I stayed throughout the centuries. My strength was in the fact that they did not know I was the guardian. If they conquered me they would conquer the land, but they didn't know that, so the land was safe."

A journey or procession usually seemed to lead to a specific place. In one of my experiences in this posture, I was taken to a palace or temple made of stone and dressed all in white. There was a row of women in this posture, all the same height, size, and age, all of whom greeted me. Upon reflection, I recognize that the shawabty figures were doing their job of welcoming me to the Realm of the Dead and making themselves available to me. Janis was in a dark room in a tomb, illuminated by colored lights in her head and spine.

John asked, "What is this posture for?" He was told that it would be a guide to knowledge, presumably about dying. Sometimes people received clear messages. Linda was told, "I am to see beauty." Michele was instructed that the crossed wrists are a reminder to create balance between male and female, masculine and feminine energies, that right over left was a greeting, expressing deference (as a servant might), while left over right conveyed aggression. The position of the body in death was considered an important aspect of preparation for the journey to the Underworld.

A sense of the continuity of life is revealed in one woman's trance. "We go toward a clearing that has a fire in the center. She [a turtle] takes me counterclockwise around it to a seat I am to sit on. . . . I walk around the fire. I see a large dipper from the moon that pours water on the fire. The fire continues to burn. The water becomes a silver stream that flows back to the sea." She remains, "grounded, serene, centered."

Performing the Posture: Stand with your knees locked and your feet close together. Make fists with both hands and cross them at the wrists in front of your body just below your breasts. Your right hand should be on top of your left hand. Keep your arms stiff and hold your elbows away from your body. Close your eyes and mouth, and face forward.

Fig. 9.4. The Shawabty Posture

Notes

Introduction

1. Belinda Gore, *Ecstatic Body Postures: An Alternate Reality Workbook* (Rochester, Vt.: Bear & Company, 1995).
2. Felicitas D. Goodman, *Where the Spirits Ride the Wind: Trance Journeys and Other Ecstatic Experiences* (Bloomington: Indiana University Press, 1990).
3. Carlos Castaneda, *The Art of Dreaming* (New York: HarperCollins, 1993), 144.
4. Rupert Sheldrake, *The Presence of the Past: Morphic Resonance and the Habits of Nature* (New York: Times Books, 1988).
5. Felicitas D. Goodman, *My Last Forty Days: A Visionary Journey among the Pueblo Spirits* (Bloomington: Indiana University Press, 1997).

Chapter 1. Living Ecstasy

1. Brian Swimme, "The Religion of the Ad," *The Sun Magazine* 305 (2001): 5.
2. Leonard Shlain, *The Alphabet Versus the Goddess: The Conflict Between Words and Images* (New York: Viking Penguin, 1998).
3. Daniel Pink, *The Whole New Mind: Moving from the Information Age to the Conceptual Age* (New York: Penguin Group, 2005); and Daniel Goleman, *Social Intelligence: The New Science of Human Relationships* (New York: Bantam Dell, 2006).
4. Shlain, *The Alphabet Versus the Goddess*, 19.
5. Robert Bly and Marion Woodman, *The Maiden King: The Reunion of Masculine and Feminine* (New York: Henry Holt and Company, 1998), 122.
6. Karl Fisch, "Did You Know: Shift Happens," *YouTube,* February 8, 2007.

Chapter 2. Defining Ecstatic Trance

1. V. F. Emerson, "Can Belief Systems Influence Behavior? Some Implications of Research on Meditation," *R.M. Bucke Memorial Society Newsletter* 5 (1972), 20–32.

2. Goodman, *Where the Spirits Ride the Wind*, 23.

3. Anna-Leena Kiikala and Mihaly Hoppal, "Studies on Shamanism," in *Ethnologica Uralica* (Helsinki: Finnish Anthropological Society, 1998), 197–220.

4. Linda Schele and David Freidel, *A Forest of Kings: The Untold Story of the Ancient Maya* (New York: William Morrow and Company, 1990), 44.

5. Barbara Tedlock, *The Woman in the Shaman's Body: Reclaiming the Feminine in Religion and Medicine* (New York: Bantam Dell, 2005).

6. Elizabeth Roberts and Elias Amidon, eds., Tewa Pueblo Prayer, in *Earth Prayers from Around the World: 365 Prayers, Poems, and Invocations for Honoring the Earth* (New York: HarperCollins, 1991), 137.

Chapter 3. Why We Use Ecstatic Body Postures

1. Ingrid Mueller, unpublished research, University of Freiburg, Germany.

2. Jeremy Taylor, *Where People Fly and Water Runs Uphill: Using Dreams to Tap the Wisdom of the Unconscious* (New York: Warner Books, 1992), 34.

3. Carlos Castaneda, *The Active Side of Infinity* (New York: HarperCollins, 1998), 189–96.

Chapter 4. The Basic
Method for Entering Ecstatic Trance

1. T. S. Eliot, "Burnt Norton," *Four Quartets* (London: Faber & Faber, 1952), 5.

2. Gore, *Ecstatic Body Postures*, 41–277.

3. Nana Nauwald and Felicitas D. Goodman, *Ekstatische Trance: Rituelle Korperhaltungen und Ekstatische Trance, Das Arbeitsbuch* [Ecstatic Trance: Ritual Body Postures and Ecstatic Trance, The Workbook] (Havelte, Holland: Binkey Kok, 2004), 85–209.

4. Malcolm Gladwell, *Blink: The Power of Thinking Without Thinking* (New York: Little, Brown, 2005).

5. Goodman, *Where the Spirits Ride the Wind*, 55.

6. Daniel Siegel, *The Developing Mind: How Relationships and the Brain Interact to Shape Who We Are* (New York: The Guilford Press, 1999), 41.

7. Juergen W. Kremer and Stanley Krippner, "Trance Postures," *Re-vision* 16, no. 4 (1994), 173–82.

Chapter 5. Healing Postures

1. For more on the Bear Spirit, see Gore, *Ecstatic Body Postures,* 49–54.
2. Ibid., 68–73.
3. Ibid., 60–67, 167–72.

Chapter 6. Divination Postures

1. Roberta H. Markman and Peter T. Markman, *The Flayed God: The Mythology of Mesoamerica* (San Francisco: HarperSanFrancisco, 1992), 9.

Chapter 7. Metamorphosis Postures

1. Princeton University Art Museum, *The Olmec World: Ritual and Rulership* (Princeton, N.J.: Princeton University, 1996), 328.

Chapter 8. Spirit Journey Postures

1. Tom Kenyon, *The Magdalen Manuscript* (Louisville, Colo.: Sounds True, 2006), 315.

2. Marija Gimbutas, *The Language of the Goddess: Unearthing the Hidden Symbols of Western Civilization* (San Francisco: HarperSanFrancisco, 1991), 150.

Chapter 9. Initiation Postures

1. Kathleen Dowling Singh, *The Grace in Dying: How We Are Transformed Spiritually As We Die* (San Francisco: HarperSanFrancisco, 1998), 2.

2. Sylvia Perera, *Descent to the Goddess: A Way of Initiation for Women* (Toronto: Inner City Books, 1981), 9.

3. Robert Masters, *The Goddess Sekhmet: The Way of the Five Bodies* (Warwick, N.Y.: Amity House, 1988), 42.

Bibliography

Abram, David. *The Spell of the Sensuous: Perception and Language in a More Than Human World.* New York: Vintage Books, 1996.

Aldhouse-Green, Miranda, and Stephen Aldhouse-Green. *The Quest for the Shaman.* London: Thames & Hudson, 2005.

Allen, Paula Gunn. *Grandmothers of the Light: A Medicine Woman's Sourcebook.* Boston: Beacon Press, 1991.

Bly, Robert, and Marion Woodman. *The Maiden King: The Reunion of Masculine and Feminine.* New York: Henry Holt and Company, 1998.

Campbell, Joseph. *Historical Atlas of World Mythology, Vol. II: The Way of the Seeded Earth, Part 2: Mythologies of the Primitive Planters: The Northern Americas.* New York: Harper & Row, 1989.

———. *Historical Atlas of World Mythology, Vol. II: The Way of the Seeded Earth, Part 3: Mythologies of the Primitive Planters: The Middle and Southern Americas.* New York: Harper & Row, 1989.

Castaneda, Carlos. *The Active Side of Infinity.* New York: HarperCollins, 1998.

———. *The Art of Dreaming.* New York: HarperCollins Publishers, 1993.

Clow, Barbara Hand. *Catastrophobia: The Truth Behind Earth Changes in the Coming Age of Light.* Rochester, Vt.: Inner Traditions/Bear & Company, 2001.

Deloria, Vine, Jr. *God Is Red: A Native View of Religion.* Golden, Colo.: Fulcrum, 1994.

Gimbutas, Marija. *The Language of the Goddess: Unearthing the Hidden Symbols of Western Civilization.* San Francisco: HarperSanFrancisco, 1991.

Gladwell, Malcolm. *Blink: The Power of Thinking Without Thinking.* New York: Little, Brown, 2005.

Goleman, Daniel. *Social Intelligence: The New Science of Human Relationships*. New York: Bantam Dell, 2006.

Goodman, Felicitas D. "Body Posture and the Religious Altered State of Consciousness: An Experimental Investigation." *Journal of Humanistic Psychology* 26 (1986): 81–118.

————. *Ecstasy, Ritual, and Alternate Reality: Religion in a Pluralistic World*. Bloomington: Indiana University Press, 1988.

————. *My Last Forty Days: A Visionary Journey Among the Pueblo Spirits*. Bloomington: Indiana University Press, 1997.

————. *Where the Spirits Ride the Wind: Trance Journeys and Other Ecstatic Experiences*. Bloomington: Indiana University Press, 1990.

Gore, Belinda. *Ecstatic Body Postures: An Alternate Reality Workbook*. Santa Fe: Bear & Company, 1995.

Hoffmann, Glynda-Lee. *The Secret Dowry of Eve: Woman's Role in the Development of Consciousness*. Rochester, Vt.: Park Street Press, 2003.

Irwin, Lee. *The Dream Seekers: Native American Visionary Traditions of the Great Plains*. Norman, Okla.: University of Oklahoma Press, 1994.

Katz, Richard. *Boiling Energy: Community Healing Among the Kalahari K'ung*. Cambridge, Mass.: Shambhala, 1987.

Leonard, Linda. *Creation's Heartbeat: Following the Reindeer Spirit*. New York: Bantam Books, 1995.

Markman, Roberta, and Peter T. Markman. *The Flayed God: The Mythology of Mesoamerica*. San Francisco: HarperSanFrancisco, 1992.

Masters, Robert. *The Goddess Sekhmet: The Way of the Five Bodies*. Warwick, N.Y.: Amity House, 1988.

Narby, Jeremy. *The Cosmic Serpent: DNA and the Origins of Knowledge*. New York: Jeremy P. Tarcher/Putnam, 1998.

Nauwald, Nana, and Felicitas D. Goodman. *Ekstatische Trance: Rituelle Korperhaltungen und Ekstatische Trance, Das Arbeitsbuch* [Ecstatic Trance: Ritual Body Postures and Ecstatic Trance, The Workbook]. Havelte, Holland: Binkey Kok, 2004.

Nelson, Richard K. *Make Prayers to the Raven: A Koyukon View of the Northern Forest*. Chicago: University of Chicago Press, 1983.

Pearce, Joseph Chilton. *The Biology of Transcendence: A Blueprint of the Human Spirit*. Rochester, Vt.: Park Street Press, 2002.

Perera, Sylvia. *Descent to the Goddess: A Way of Initiation for Women*. Toronto: Inner City Books, 1981.

Pink, Daniel. *The Whole New Mind: Moving from the Information Age to the Conceptual Age*. New York: Penguin Group, 2005.

Princeton University Art Museum. *The Olmec World: Ritual and Rulership*. Princeton, N.J.: Princeton University, 1996.

Rockwell, David. *Giving Voice to Bear: North American Indian Myths, Rituals, and Images of the Bear*. Niwot, Colo.: Roberts Rinehart, 1991.

Schele, Linda. *Hidden Faces of the Maya*. Poway, Calif.: Alti, 1997.

Schele, Linda, and David Freidel. *A Forest of Kings: The Untold Story of the Ancient Maya*. New York: William Morrow and Company, 1990.

Sheldrake, Rupert. *The Presence of the Past: Morphic Resonance and the Habits of Nature*. New York: Times Books, 1988.

Shlain, Leonard. *The Alphabet Versus the Goddess: The Conflict Between Words and Images*. New York: Viking Penguin, 1998.

Siegel, Daniel. *The Developing Mind: How Relationships and the Brain Interact to Shape Who We Are*. New York: The Guilford Press, 1999.

Taylor, Jeremy. *Where People Fly and Water Runs Uphill: Using Dreams to Tape the Wisdom of the Unconscious*. New York: Warner Books, 1992.

Tedlock, Barbara. *The Woman in the Shaman's Body: Reclaiming the Feminine in Religion and Medicine*. New York: Bantam Dell, 2005.

Walsh, Roger N. *The Spirit of Shamanism*. New York: G.P. Putnam's Sons, 1990.

Wilshire, Bruce. *Wild Hunger: The Primal Roots of Modern Addiction*. Lanham, Mass.: Rowman & Littlefield, 1998.

Wolf, Fred Alan. *The Dreaming Universe*. New York: Simon & Schuster, 1994.

Ywahoo, Dhyani. *Voices of Our Ancestors: Cherokee Teachings from the Wisdom Fire*. Boston: Shambhala, 1987.

How to Use the CD and Ritual Postures to Enter Ecstatic Trance

The CD included with this book has four 15-minute segments. You can use any of these for your ecstatic trance session.

First, decide which of the four rattling and/or drumming segments you want to use and cue up the track so that all you have to do is turn it on.

If you are alone, you may want to use earphones to exclude all of the ambient sound and intensify the sound of the rattle or drum. If you choose to use earphones, have them ready to put on. If you are working in a small group, have the sound level on the speakers set at the right volume.

Begin the ritual just as described in chapter 4, "The Basic Method for Entering Ecstatic Trance":

1. Decide which posture you plan to use and practice it so you can assume the posture easily.
2. Set up an altar in the center of the space you will be using.
3. Have smudge, such as a sage or cedar smudge stick, ready and once it is lit, allow all who are participating to smudge themselves in the smoke. Then extinguish the smudge stick.
4. Set the sacred space by calling on the spirits of the four directions (east, south, west, north) and the Sky Beings and the Earth Spirits, using a rattle, drum, or other instrument, including song.
5. Feed the spirits you have called with corn meal, tobacco, or other herb.
6. Do the breathing exercise as described on page 36 for 50 breaths or about 5 minutes.
7. Put on your earphones if you are using them.
8. Assume the posture.
9. Turn on the CD.
10. At the end of the 15-minute session, first turn off the CD; then take a minute or two to make the transition back to ordinary consciousness. Then record your experience, and if you are in a group, share your experiences with each other.